Flights of Adventure

Ken Larson and Tom Holton

Copyright © 2009 by Ken Larson and Tom Holton

NorLightsPress.com
2721 Tulip Tree Rd.
Nashville, IN 47448 U.S.A.

All rights reserved. No part of this book may be reproduced or transmitted in any form or by any means, electronic or mechanical, including photocopying, recording, or by any information storage and retrieval system, without written permission from the author, except for the inclusion of brief quotations in a review.

Printed in the United States of America
ISBN: 978-1-935254-17-1

Book Design by Nadene Carter
Cover Design by Vorris "Dee" Justesen

First printing, 2009

Flights of Adventure

Ken Larson and Tom Holton

Dedication

Flights of Adventure is dedicated to the aviators, dreamers, and adventurers who came before us, and upon whom our own dreams and adventures are founded. Thank you all.

Table of Contents

Introduction 1

Blind Approach, by Tom Holton 4

The WASPS of World War II, by Marion Stegeman Hodgson 19

GOLMA, by Ken Larson 31

Two-Fer Day, by Kevin Kasberg 56

The Milgram Experiment, by Ken Yamada 71

Walking in a Hurricane, by Carol Pilon 107

Chasing Pablo, by Nick Qualantone 128

The Ice, by Paul Derocher, Captain, USN, Retired 143

Any Landing You Can Walk Away From, by F. Victor Sullivan 178

Reluctantly Airborne, by Olive L. Sullivan 192

Biographies 196

Introduction

The romance of flight is embedded in the human psyche just as surely as the need for food and drink. Ancient myths remind us of this, but we feel it most strongly when our hearts soar as a bird takes wing.

I recently considered this thought while lying on my back in the sun, watching cormorants approach the lake near my home, with their wings arched and unmoving like the stiff metal wings of a Cessna. Their necks were arrow straight, their attention focused totally ahead. These birds reminded me of fighter jets flying initial patterns toward a waiting aircraft carrier. The glides were long, graceful, and fast, as one bird after another came in overhead. I noticed my rising pulse rate and felt an ancient need to be like them. I wanted to fly. I wanted to stretch my arms and feel the rush of a high-speed descending glide. I wanted to rise in the thermals and circle as I climbed higher, leaving the world behind, thinking only of the uncertainty lying ahead.

My thoughts were interrupted by an unusual sound—a mix of whistling and soft cries. Swans! Wild swans coming in from the north, still very high, yet starting a formation turn to the east toward rocky cliffs. I clambered to my feet and ran to the water's edge. Concealing myself among the dry cattails and long brown grass, I knelt in shallow, icy water and silently watched.

Nine swans flew high over the small lake, in perfect chevron formation, precisely matched in speed, each identical to the next. As if on signal from the leader, each bird's neck arched back, the white and black heads rising and bending aft almost above their backs. All wings in the flock curved downward. Their legs extended at the same time, like the mechanical landing gear on huge airplanes or speed brakes on supersonic fighters. The swans descended almost vertically, coming straight toward me. I could see their eyes, their glossy black bills, and hear the rustling of wing tip feathers and black feet making small adjustments as they dropped from the sky. I held my breath: it seemed they would smash into the lake in a jumble of white impacting blue water. But again, as if on cue, every magnificent bird raised its feet, lowered its neck, and leveled off, trading speed for forward momentum, breaking their fall toward the water. A few feet above splashdown, the birds—in perfect unison—lowered their landing gear, put down their flaps, and smoothly, silently touched down, each shiny black foot making a small wake as they glided to a stop.

I was breathless, having witnessed one of the most incredible displays of flight I've ever seen. I wondered about the first human who witnessed such an event, and whether he or she felt the urge to leave the earth and soar into the heavens.

Our myths tell us that long-ago witnesses did imagine human flight. We thrill to stories of adventure: the romance of the Wright Brothers, Amelia Earhart, Charles Lindbergh, and even John Glenn. They were the famous flyers, but they weren't the first to seek and find adventure in danger and death-defying feats, nor were they the best. Find a group of pilots at any gathering, and you'll hear hair-raising stories of derring-do, adventure, and everyday heroism. What would motivate a seemingly normal Canadian woman to climb out of a perfectly good airplane and do gymnastics on its wing while the plane does aerobatic maneuvers? Why would a young woman in the 1940s give up the comfort of home and join a cadre of females who would become aviation pioneers, risking their lives, putting up with nearly intolerable conditions, and facing the scorn of many of their male counterparts? What possesses a pilot to risk his own life to rescue a stranger, or another to fly in the most dangerous, coldest places on earth? Why are some pilots capable of solving life-and-death problems when others might panic or give up?

These are the stories you'll read in **Flights of Adventure.**

This book is a compilation of true stories from the world of aviation as told by the pilots, performers, and passengers who lived them. Even though the book revolves around aviation, its appeal is universal. Anyone who loves to travel or has a sense of adventure (even the armchair type) will find these stories exciting. Anyone who's watched in fascination as birds fly or clouds glide past—anyone who dreams of faraway places and interesting people—will find much to like in these pages.

Sometimes humorous, sometimes dangerous, but always great reading, **Flights of Adventure** takes us on a journey around the world. From the coldest place on earth at the South Pole to the intense heat of the East African deserts, then across North America, these stories bring to life the excitement, danger, and fun shared by those who fly, or just dream of it.

Most of the stories are written by pilots, but non-flyers also weigh in. Each chapter is a stand-alone tale, taking you to the heights of adventure. The stories are touching, humorous, exciting, and often dangerous or miraculous. We hope you enjoy reading them as much as we enjoyed the telling.

BLIND APPROACH

by Thomas Holton

Flying is hours and hours of boredom sprinkled with a few seconds of sheer terror.

~~Gregory "Pappy" Boyington
1912-1988

On Saturday, April 16, 1977, I was up early for my morning run along the bike/walking trail in the Kingwood section of Humble, Texas. My wife, Anne, and I lived in a second-story apartment in a newly developed planned community. I stood on the balcony with a cup of coffee, taking in the cool spring air and looking over a stand of loblolly pine just off the balcony. The pines were new growth trees, planted to buffer a new housing area across the bike path, so the tops of the trees were even with our balcony floor. The sun was up, the air still, and the sky clear blue *sans* clouds. This would be a lovely spring day.

Then the phone rang.

Anne answered the call. "Tom, it's for you, scheduling." I shook my head vigorously as she handed me the phone.

"Tom, here." I had a bad feeling about this call.

"Tom, this is Chris in scheduling, and we have a charter run from Houston to Peru and back to Houston for the reserve crew. You have a 1600 local time show."

My heart increased several beats per minute; I knew it would be a long time before I'd see a bed and sleep. It was only nine o'clock in the morning, and I felt an edge of uncertainty.

I was a first officer on the Lockheed Electra for Transamerica Airlines, headquartered in Oakland, California. The Lockheed Electra is designated an L-188, a four-engine turboprop that cruises in the altitude range of 19,000 to 24,000 feet. The military version is the Navy P-3 Orion aircraft. Transamerica had nine L-188s operating as freighters. The planes were passenger aircraft operated by KLM in Europe when they were new, but they ended up in the USA and became freighters as they aged.

At 3:30 in the afternoon, with bags packed, I kissed Anne goodbye and headed for the Houston Intercontinental Airport. Anne knew I wouldn't return until at least 6:00 in the evening on Sunday, assuming everything went as scheduled. As I drove west on Kingwood Drive to Highway 59 for the short drive south to Houston Airport, the sky was still deep blue with only small puffy white clouds. On such a beautiful day, it was hard not to feel good, but something still kept me off balance; an uncomfortable feeling nibbled at the edges of my brain, starting at the base of my spine like a thorny little creature. I tried unsuccessfully to shake it, hoping to enjoy the day. Instead, I kept reviewing the flight plan: we would carry a load of big pipes from Houston to Iquitos with a fuel stop in Panama. When they told me Iquitos was located on the east edge of Peru on the Amazon River, my stomach tightened a little more. Airports in such locations usually don't maintain high standards. They can be challenging on the best of days.

After parking at the cargo area, I saw Captain Paul Anderson* and Flight Engineer Earl Butler* had already arrived. Paul lived in the Ann Arbor area and commuted to Houston, so he'd been staying in a hotel on "his dime" and had checked out early to avoid paying an extra night. Earl lived in San Antonio and drove to Houston. Bob Fallon*, the loadmaster, worked at loading a pile of long, black pipes into the L-188. Each pipe was about 20 feet long and four inches wide—quite a load for this old bird.

Paul was just shy of six feet, thin and in good shape, though prematurely gray. Cool and confident under fire, he seldom showed emotion. Having served on the L-188 for a long time, Paul was respected for his airmanship and ability to handle a crew.

* NOTE: Author did not use these individuals' real names.

Earl, a retired air force engineer, was also known for professionalism, bringing strict discipline learned in the Air Force to his job. He knew the aircraft systems and followed operating procedures to the letter. Earl had a barrel chest and the build of a wrestler, with graying temples and a confident smile.

I was new to the airplane, but I had capable crewmembers with whom to face the jungles of Peru. The edgy feeling dissipated a bit as I worked with my new crew.

"Good to see you—hope you got plenty of sleep," Paul greeted me.

"Just the usual," I said. Obviously, he wasn't looking forward to being up for 30 hours, either.

"Hi, Earl. Nice to see you again," I said, as I shook Earl's hand. We'd flown together just three weeks earlier.

"Are you up for this?" he asked.

"Never ready for an all-nighter, since I've already been awake since 6:00 this morning," I said. It seemed I wasn't the only guy on edge about an all night trip to the upper Amazon.

Bob was loading the pipes in the plane through the side cargo door as I approached the aircraft. After stowing my bags, I began the pre-flight duties, checking navigation charts for all the airports and alternate airports to make sure all charts were on board and up to date. I noticed the chart for Iquitos had an asterisk by the tower frequency, indicating they didn't operate 24 hours a day. Hours of operation weren't indicated on the chart. I ran a scenario in my head, considering the impact of a closed tower. Leaving Houston on Saturday evening would put us into Iquitos around 6:05 a.m. on Sunday morning. I went into the office and found Paul on the phone, filing our flight plan to Panama.

"Paul, the tower isn't 24/7," I informed him.

Paul nodded, tucking the phone under his chin and cupping a hand over the mouthpiece. "I'll call dispatch in Oakland and see if they know the hours." He gave me an understanding thumbs-up and returned to the phone call.

A company dispatcher in Oakland, California created our flight plans, computed the fuel burn for the flight, checked weather, and then faxed the information to Paul in Houston. Our dispatcher assured Paul the tower at Iquitos would open at 6 a.m., and the weather forecast called for clear skies. No problem.

Shortly after 5:00 in the afternoon, our final pre-flight duties were almost complete. Bob made sure the pipes were secure. Earl finished the pre-flight systems inspection, and we had the plane fueled when Paul returned to the plane.

"The airport only has a VOR approach, if we need one," I told the captain.

"If we depart at 6:00 this evening, with the one hour ground time in Panama, we should arrive around 6:10 in the morning at Iquitos, based on the three-hour leg from Panama to Iquitos. Dispatch confirmed the tower opens at oh-six-hundred," Paul announced. "I asked the dispatcher to call Iquitos and confirm the tower opens at 6:00 in the morning."

At 6:16 p.m. we headed to the end of the runway to take off toward the west, into the dusk. The tower cleared us for takeoff, and Paul called for "max power," with Earl setting the four thrust levers to maximum torque on the Allison 501-D-13 engines. We lifted off the runway into the calm south Texas evening. Paul made the takeoff. I would fly the next leg into Peru.

The L-188 didn't have navigation or communications radios for oceanic flying, which prevented us from turning south over the Gulf of Mexico for Panama. We headed along the coast of Texas and into Mexico using the land route. The air was smooth, the stars bright, and the lights of cities and villages glided beneath us throughout the seven hours to Panama. When we landed there at 2:00 in the morning, the thunderstorms that occur every day had dissipated, leaving only small, fluffy, cumulus clouds.

After shutting down the engines, Paul turned to us. "Stretch your legs; I'll file the flight plan, check the weather, and pay the handler. Double check the fuel loading, okay?"

Thirty minutes later, Paul briefed us and handed me the weather charts to review: "Iquitos is still forecast for clear, but there are thunderstorms all across the Andes."

"We're legal on fuel, based on the forecast and dispatch numbers," I said, checking our log. We had enough fuel to fly to Iquitos and our alternate airport in Quito, Ecuador, plus the required reserve of about an hour more.

"Let's go, then." Paul looked each of us in the eye, giving us plenty of opportunity to object if we saw something wrong. "Oh, yeah, Tom. I checked in with HQ again. You know … just in case. Dispatch says the airport in Peru will definitely be open at oh-six-hundred. No sweat."

It felt good to have the controls as I taxied the big plane to the runway. We worked as a smooth team, and I was happy with this crew. I banked the airplane south toward the middle of the Amazon jungle at the east end of Peru. I had to climb higher than expected because of the mountains, making our ground speed slower than the plan and burning more fuel. The wind was also different from the forecast, and we began falling behind the flight plan in fuel and time. During the last few hours of darkness before sunrise, a little over an hour since our departure from Panama, we approached the Andes Mountains. I was surprised to encounter high, building clouds at this time of night—especially the monsters we saw in front of us, which were thick and full of moisture.

"I guess we need the engine heat on, Earl," Paul said.

"You got it—just hope we don't need it for long," Earl replied. "The radar shows more thunderstorms than were forecast, so we may be in the clouds awhile."

"We're ten minutes behind our flight plan and ten minutes of reserve fuel is gone," I commented. *Already gone,* yet we still had a long way to fly. Engine heat would control ice on the props and wings, but it would cause us to burn extra fuel. That itchy feeling of uncertainty returned.

Our weather radar indicated the thunderstorms were not scattered as forecast; they were everywhere, and massive. Our radar painted the thunderstorms with great clarity, and the flashes of lightning ahead seemed like warnings not to enter. To avoid these killers, we had to turn 45 to 60 degrees off course and fly around them. We'd fly 40 miles west of one storm, return to the airway, and then fly 40 to 50 miles east to squeeze past the next one, adding distance to our trip. Our reserve fuel fell lower.

"Even crossing the Midwest in the spring, I've never seen so much St. Elmo's fire as we have tonight," Paul said.

Small bolts of electric-blue gremlins danced on the windshield, the light show almost hypnotizing. I touched the windshield and felt the tingle of electricity in my hand; all the hair on my arm stood at attention.

"I hope this ends soon. We don't need any excitement tonight."

In the pre-dawn hours, I assessed our situation: our mileage from Panama City to Iquitos was way above the plan, engine heat was on nearly the whole time, the fuel supply showed well behind at each checkpoint, the weather was much worse than forecasted, and we'd all been awake for 20 hours. Not to mention, stress played havoc with our nerves.

Colonel Francisco Secada Vignette International Airport in Iquitos, Peru, is located three degrees south of the equator. Near the equator, the amount of daylight and darkness is close to equal: 12 hours each day all year long. At Iquitos, sunrise would be shortly after 5:30 in the morning. The sun would be an hour above the horizon when we landed at 6:30.

Iquitos sits at the confluence of three rivers—the Nanay, the Itaya, and the Amazon—and is the largest city in the Peruvian rainforest. Approximately 300 feet above sea level, it sits about 1,000 miles from the point where the Amazon River empties into the Atlantic Ocean. At the time, Iquitos was the largest city in the world that couldn't be reached by road. The nearest road was at Nauta, about 30 miles south.

The sky to the east turned bright pink with the approaching sunrise. The air was now clear and smooth, with only an hour to Iquitos. We began planning our arrival and approach, feeling a little better as our destination seemed within our grasp.

The review of the descent and the type of approach were important on this international charter flight, because none of us had ever been to this airport. From our instrument approach charts, we knew Iquitos had only one instrument approach—a non-precision VOR to guide us into the runway if clouds were present. The VOR is a reliable system that allows an aircraft to descend through clouds to around 400 to 500 feet above the ground, a point where the pilot can see the runway and still have time to clear difficult terrain and obstacles. If the cloud layer is lower than 500 feet, the procedure is of little or no use in landing.

We spent the last twenty minutes of flight descending to the airport from our cruise altitude of nearly 24,000 feet. Though we expected clear weather, I reviewed the VOR approach, noting the bearings and courses to fly, and especially looking at the safe altitudes.

An airport's VOR station emits a radio signal in a straight line, which we expected to receive on our cockpit instruments at around 23,000 feet of altitude. Surely, we'd get the signal by 150 miles out. I was feeling much more comfortable, knowing we had clear weather and a backup instrument approach.

A few minutes after 6:00 and well inside the 150-mile range from the airport, we still hadn't received the navigation signals.

"Probably a low-altitude, low-power transmitter for approaches only. We'll get it in a few miles," Paul assured us.

I hoped he was right. We double-checked notices to airmen that would advise if the station was inoperable. Nothing. We waited a couple more minutes and at 6:15 made a call to the control tower.

"Good morning, Iquitos tower. This is Transamerica 852," Paul transmitted.

No answer.

"Iquitos tower, Transamerica 852 on 118.1. How do you read?"

Still no answer.

"I hope the controller decided to come to work," I said, glancing again at the instrument approach plate.

"It is Sunday morning; hope he didn't sleep in," Earl said.

"We may be on our own, boys," Paul added, half-heartedly. "Any of you speak Spanish?"

That itchy little creature started bothering me again. This time, it felt large and mean—gnawing at my spine, beginning to form knots in my stomach. None of us said a word, but we all felt the same thing: This felt wrong.

At 60 to 70 miles northwest of the airport, we needed to descend further, but we couldn't raise anyone at the tower and had no navigational aids. We'd been cleared to descend by the en route controller, but didn't trust that entirely because of difficulties with the language. We'd all heard stories of fatal plane crashes where the controller didn't understand the pilots and conveyed bad instructions. Without accurate navigation aids, we didn't have an exact position—an issue that might prove deadly in this terrain, in an isolated part of the Amazon drainage.

I descended further. We had no choice.

At the 60-mile range, we saw the general area of the airport through the morning haze. Iquitos truly was perched on the edge of the Amazon

River. All of us stared ahead, straining to find the airport. We couldn't see it. Instead, as I descended below 10,000 feet, the sight of the area around the town and airport caused that spiny little creature in my backbone to clamp down, and the queasy feeling in my stomach became a solid knot. A dense blanket of fog covered the area over the Amazon River, extending out five or six miles. The fog was no more than 200 to 300 feet thick, but it hugged the river and the ground, totally obscuring the town and the airport.

We were in deep trouble.

Fog over the Amazon River had a different cloud top form than fog over land. Over the water, the top of the fog was around 100 feet, thick and smooth. Over the jungle, town, and airport, the fog was around 200 to 300 feet thick, and a bit lumpy. The difference in the tops occurred because of temperature variations. The water was a constant temperature, so fog clung to the river as though holding on for dear life. Over the landmass, the earth was heating from the sunlight, which caused the fog to rise a bit. Rather than dissipating, this rising effect made the fog deeper. A few holes were starting to open in the fog, with the bottom of the fog releasing its death grip on the land and slowly rising—very slowly. It was already 6:30, and we were nearly out of gas. We needed a plan of action—immediately.

Paul sensed the urgency and broke our silence. "This is the pits, boys. Do we have fuel to make Quito?"

Earl had already calculated the fuel requirements. "Barely, Paul. I'm afraid we'll run out just before we make the field." Then, anticipating Paul's next question, "Yes, I calculated with our most efficient power settings."

"Consider our options, boys," Paul said. "I don't think we have many."

Shortly after 6:30, our L-188 was passing 3,000 feet over the area of the airport. We could guess the airport's approximate location by observing the fog pattern, which formed a perfect outline of the meandering river. By judging S-turns in the Amazon River, we knew about where the airport should be below the fog. The control tower still didn't answer our radio calls. The VOR station did not transmit an identifier to indicate it was operating, nor did our navigation needle in the cockpit point to the station. We were flying blind, unable to find the runway. Our only hope of descending below the clouds depended on the VOR station transmitting properly, and it was dead—totally silent. Useless.

Paul spoke again. "I don't want to risk running out of fuel over Quito and killing a bunch of people on the ground. It's an hour away without a headwind, and we haven't been so lucky on weather, have we? I'd rather use what little fuel we have left trying to save our butts. I'd rule out Option One, flying to our alternate. You, boys?" Paul was keeping us all in the decision-making process, making us think through the situation with him.

No one argued. I also considered that one never wants to crash land an L-188 without the engines running, because the hydraulic system will quit working, as will the flight controls—leaving no way to steer the big transport plane. Having the engines stop running over a city the size of Quito would not be a good thing for us, or the people below.

The creature crawling up my spine told me we were going to crash. Our options seemed to be not *if* we crashed, but *where* to crash. We needed to do it intentionally before we ran out of fuel.

Earl spoke next. "Crocodiles and piranha, or huge trees and jaguars?"

Bob said, "Ditch into the river and get eaten by crocodiles or piranha. This is no way to spend a Sunday, if you ask me."

"If we take a gradual descent into the jungle, away from town a bit, the trees might absorb our energy; we might live through it. I'd rather go for the trees than the Amazon," Earl added. "Of course, that's where the jaguars live and dine."

Paul looked at me, expecting a wisecrack or some lame-headed comment, but I focused on flying the lumbering plane, occasionally glancing at the ground.

"Blind approach," I said. "Just trying a controlled descent onto the ground close to the airport area is an option, but it puts people on the ground at risk. I doubt we can walk away from hitting huge trees in the jungle, and I agree being fish food would ruin our whole day. How much time do we have?"

Earl added, "Coming down over town and hitting any building will have the same results as smacking a tree out there, except for people in the building."

I chimed in, "And in the river, we'll get out of the plane because it will float for a while, but with no raft…" I dropped the subject.

Paul turned to look at us. "I'll try the controller who cleared us for the descent and see if he knows why the tower and VOR aren't operating."

No answer, probably because we were too low for reception.

"Do you think you can hit the airport in a slow descent?" Earl asked me.

I thought about the blind approach as I flew the plane at 3,000 feet across the top of where we suspected the airport to be, based on a curve in the river. I turned us toward the southeast, and then made a turn to the left after passing the Amazon River. The town and airport were on the west side of the river. As each minute ticked by, the fog slowly rose because the earth was warming. Fog still clung to the water, which made the river more pronounced than when we first approached the city. As I headed back north, Paul was still calling the control tower, hoping for a response. Then an idea started coming into focus for me.

I asked Earl to remove the WAC—World Aeronautical Chart—from the navigation case. These maps are used for visual reference of the world surface as if looking down from space, showing details of terrain, cities, and things like roads, railroads, power lines, and airports. I figured the one for this part of Peru might show an area around Iquitos with a clear space void of trees and population. Glancing at the WAC, I noticed a radio tower on the edge of town northeast of the airport—a tower for an AM radio station, with the station's frequency printed on the map.

The L-188 had two ADF receivers that could pick up the signal for an AM radio station. Because this was a civilian radio station—the kind you tune in from your car radio—and not maintained by the Aeronautical Department of Peru, it had no published approach procedure to line us up with the runway to descend safely. There was, in fact, no information at all except the transmitter tower's location, shown on the map by an upside-down "V" and its frequency—950 on your radio dial, or something like that. However, I thought of a scheme that would use the AM station to pinpoint our location. The clock ticked past 6:35 and still no communication with the control tower, but Paul kept calling.

"Paul," I said, "let's create our own approach off the AM station. See if we can get down a little and see the ground. Blind approach, but with a little help."

We were out of options, out of time, and very nearly out of fuel. No one objected.

I dialed the radio station frequency into our receiver. Music and then the announcer, doing his morning show in Spanish, came into the headset. I turned west and flew for one minute, then turned south and flew another minute. The ADF needle on the cockpit instrument pointed at the radio station even as I made the turns. We could fly to the radio station following the needle on our instrument, and we knew that location because of the map. If we could have seen the ground, we'd have spotted the radio transmitter tower below as we watched the needle on the instrument panel swing from pointing up, toward the nose of our plane, to pointing down on the instrument dial, toward our tail.

The fog lifted a little higher above the ground, and small, ragged holes became visible. We saw tiny patches of ground through the holes, but the bottom of the clouds was still less than 200 feet above the ground where the fog transformed into puffy cumulus clouds.

After eyeballing the map and drawing a mental line from the radio station to the airport, I suggested to Paul, "If we flew over the radio station and then tracked out from the station with a bearing of 220 degrees, we should pass somewhere near the end of the runway. Want to try it? We'll have to descend into the clouds to see the runway, and we have no idea what obstacles might be near the airport."

Paul nodded. I flew directly over the radio station, turning to keep the needle on our instrument pointed toward the top of the dial, toward the nose of our plane in the direction we were flying.

We all agreed it might work. Since I'd been flying the plane and Paul was making the radio reports, I asked if he'd prefer to fly, since he was the captain, and I would take over the radios. I asked this because he would be the main focus of the investigation if this didn't end with a successful landing.

Paul replied, "No, you fly. I'll look for the runway."

Maybe he thought the hardest part was looking for the runway. We only had enough fuel for a couple of approaches, and then it would be time to set up for a controlled crash.

I fought the knot in my stomach, forcing total concentration on flying. To think about the possible outcome and worst case scenarios would surely lead to disaster. I didn't know how Paul and Earl felt, but I thought we had a slim chance. I was in my early thirties, and I wanted to go home. My life and career were NOT going to end in the jungles of Peru. My wife was in Houston. I would be back Sunday night. Our crew had a plan, and we had a chance.

I flew over the radio station. The needle on the plane's instrument rotated, pointing the head of the needle at the tail of our plane, toward the place we'd just passed.

Paul looked at the VOR approach plate and commented, "Let's go down to 500 feet above the ground, the height the VOR would have taken us. That should keep us clear of any obstacles."

"Sounds good," Earl added.

"Just call my altitudes," I replied.

With the gear and flaps set for approach, I reduced power on the engines and started a descent. At 500 feet above the ground, I maintained the altitude as we bounced in and out of the small puffy cumulous clouds, turbulence adding to the angst. Paul was looking outside and as straight down as possible, hoping to spot the runway.

"I saw it! We passed over the runway. Don't know if we were near the end, I couldn't see any numbers. Go around."

Paul almost sounded excited. Better excitement than fear, I thought.

"Go around power," I commanded for Earl to set.

"Positive rate," Paul called as the altimeter started to climb.

"Gear up. Leave the flaps set for approach, we'll make it a tight pattern," I said.

"The nose was about 30 degrees left of the runway heading," Paul added, as I made a hard right turn to the northeast. Climbing to 1,500 feet above the ground put us above the cloud tops. I headed straight back to the radio beacon for a second approach.

"This time we'll make the heading 210 degrees after passing the station to aim closer to the end of the runway." I paused. "Then we'll go down to 300 feet above the ground."

I felt this was it. Do or die, literally. But I announced, "We'll land on this approach, and I'll buy the beer when we get home."

Bob made a grunting sound. Paul smiled but stayed focused on the outside. This time we were in the base of the clouds, but Paul was getting short glimpses of the ground.

"I saw the numbers! The end of the runway!" Paul called with even greater excitement in his voice. "We passed right over the end of the runway. Right over the top of the damn asphalt. Go around, Tom. We're gonna make it, dammit. One more try."

Nobody asked for a check of the fuel remaining. We were quiet, giving Paul a chance to call out altitudes and the runway sighting we all hoped for.

We repeated the same procedure as the last approach, climbing to 1,500 and flying the needle back to the Spanish station. I had a fleeting thought, wondering what the guy giving his morning show on this station would think if he had any idea what was happening overhead.

"Tom, let's take it ten degrees more to the left, heading 200 degrees from the radio station." This would be our third blind approach. We knew it had to work this time, as precious fuel was being pushed through the four turboprops on each approach. Every go-around wasted huge amounts of fuel. It was now 6:55 a.m., three hours and fifty-five minutes since we departed Panama City.

"I'll stick her in the grass next to the runway if I can't hit the runway. We can't do another go-around. Okay with you, Paul?" I asked.

Paul and Earl agreed. Upon leaving the radio beacon, I flew a 200-degree heading, and we decided to take it to the limit and descend to 200 feet. This should put us below the clouds as they were slowly rising, but we risked colliding with an obstacle or terrain. If we didn't land this time, it was back to the huge trees and jaguars, or the crocodiles and piranha options. All other choices were out.

We knew the runway would be to the right of the nose, because each pass had us crossing the runway with a 20- to 30-degree angle between the nose of the aircraft and the runway heading.

Just as I started to level the aircraft at 200 feet, Paul called, "Runway at two o'clock!" and pointed off to the right of our nose.

I banked to the right and pulled the thrust levers to idle. There would be no go-around. I held the nose up, losing speed, slowly bringing the plane to the pavement. On the runway or in the grass, the plane was going to be on Mother Earth. Paul gave up calling out altitudes, and the relative silence with the engines idling was broken by the beautiful sound of tires contacting rough asphalt. We rolled down the runway at 6:58 a.m. Sunday morning. There was no cheering or yelling or celebration. The cockpit was very quiet. Earl would later tell us he calculated we had 15 minutes of fuel remaining at the most, probably less because of the amount we would burn doing a go-around at full power. We would not have survived another attempt.

The terminal building at Iquitos was a small concrete structure only 30 to 40 yards long and maybe 20 yards wide. We taxied up to the building and set the parking brake. It was now 6:59 a.m. Engines one, two, and three were shut down. Seeing a fuel truck coming, we decided to leave number four running to have electricity and keep the radios and instruments functioning while off-loading the plane, as well as to help restart the other engines.

Earl opened the cargo door, and Paul went into the building to see if he could find anyone waiting on the huge cargo of big pipes. I stayed in the right seat, monitoring the parking brake and number four engine. The time was now 7:01 a.m. I happened to notice the needle for the navigational VOR swing and point in the direction of the VOR station, and the OFF flag disappeared, indicating the station was transmitting. The VOR needle and station we'd hoped for and so desperately needed had come to life.

At that moment, I heard a Spanish accent in my headset: "*Buenos dias,* Transamerica 852, welcome to Iquitos. We are open at seven o'clock."

I took a deep breath.

While we waited on the ground, I had plenty of time to think. Why did we arrive when the tower was closed? Who was at fault? Why was the navigation radio not transmitting? Even when the control tower is closed, a VOR navigation radio normally operates 24/7. Iquitos' tower obviously shut down the VOR when the tower was closed, but likely neglected to put the information in the notices for pilots. Welcome to South American aviation.

The biggest question: Why didn't the dispatchers in Oakland plan the arrival at 7 a.m.? They told Paul in Houston the tower opened at 6 a.m., so the flight was planned for an arrival shortly after 6:00. How could they be off by one hour? Had the dispatcher even called the tower on Saturday to check the time of operation? All these questions went through my mind as we waited for the cargo to be off-loaded. If the dispatcher had indeed called the tower, he may have asked the wrong question. Then again, maybe the tower was scheduled to open at 6:00, but the controller had a late Saturday night party and overslept. Many scenarios went through my mind, but I never found the answer.

The three of us didn't discuss the landing at Iquitos on the return two legs to Houston. That problem lay behind us, and reliving it made no sense. Perhaps we were too tired. Upon leaving the aircraft on the cargo ramp in Houston, Paul hurried to a pay phone on the wall and made a call. I figured he was calling his wife to check in. A few seconds later, he hung up the phone, walked over to Earl and me, and said, "I just quit."

By the time I drove back to Kingwood late that Sunday night and climbed the stairs to our second-story apartment, Anne was already in bed. I closed the bedroom door, went to the kitchen, and poured three fingers of bourbon into a glass. I stood on the balcony with the sliding door closed and drank the Jack Daniels while gazing over the tops of loblolly pine trees and watching the stars under the Texas sky. I had to make a decision: Should I tell my wife how close we came to becoming a part of the Peruvian landscape or keep it a secret? I chose the latter. I didn't want her to worry every time I left for a trip.

The WASPs of World War II

by Marion Stegeman Hodgson

One of the best-kept secrets of WWII is the fact that the Air Force (then the Army Air Forces) trained women pilots to fly every fighter and bomber in the AAF arsenal and perform every task except combat. Years later, in the mid-1970s, the Air Force apparently forgot the service of the WASPs (Women Airforce Service Pilots). The Pentagon, with a roll of drums, announced they planned to train its first women military pilots.

We WASPs were accustomed to being left out of history books, but this was too much! We launched a campaign in Washington D.C. to have WASPs recognized as pilots who put our lives on the line when our country needed us. After all, 38 of us died in crashes. Congress responded. At last, they acknowledged our service and officially made us veterans, and the Air Force issued honorable discharges and medals.

Even so, as late as 1995, the WASPs were forgotten women. An aviation magazine ran a story indicating women didn't enter Air Force pilot training until 1976, oblivious to the strenuous training program WASPs completed in 1943 and 1944, exactly like the male cadets' program, with the exception of gunnery.

I was indignant. Having been a WASP, I had to respond in defense of the sisterhood. The magazine printed my letter, which read in part:

> How about the WASPs? We entered Air Force pilot training in 1942 during WWII. True, we did not fly combat, but 38 of us were killed in crashes in the line of duty. So it was serious piloting.

By the end of September 1944, WASPs were delivering three-fifths of all fighter planes. Besides ferrying, we towed targets, did tracking and searchlight missions, radio control flying, weather hops, administrative and utility flying, simulated strafing, and test piloting. Many gave instrument rides and refresher courses to instructors. We flew everything from primary trainers to the B-29 Superfortress, including Mustangs, Thunderbolts, B-17s, B-26s, and C-54s.

We did anything they'd let us do, knowing we had to try harder and gripe less than the men with whom we flew. The fact that they made more money and had more perks meant less than nothing to most of us. We thanked God for the chance to serve our country in a unique and exciting way.

I had the opportunity to write another letter in February, 2008, when the *London Times* printed an article by their defense editor concerning England's "unsung flying heroines." He made the following misstatement: "They [the women pilots in England who ferried Spitfires] were the only women from among the Western Allies who flew in the war."

In a letter to the defense editor, I corrected that statement, telling them, "There were over a thousand women military pilots in the United States during World War II. They were called WASPs (Women Airforce Service Pilots), and they flew everything the Army Air Forces had and performed every aerial duty except combat." As far as I knew, they never printed the correction.

The desperate need for pilots became obvious from the first day of war. In the first few months after Pearl Harbor, German submarines sank our oil tankers within sight of beaches on the East Coast. Our recruits were drilling with wooden rifles. Half of our Navy was destroyed at Pearl Harbor. Uncle Sam needed pilots. World famous aviatrix Jacqueline Cochran suggested the Army Air Forces should train women pilots for stateside service so men could be released for overseas duty.

As a result of Cochran's ambitious concept, more than a thousand girls won the Air Forces' silver wings. Some (like me) went into the Ferry

Command, picking up new planes at factories and delivering them to air bases around the country or flying old planes from base to base for repairs.

Our immediate boss in the Ferry Command was beautiful, soft-spoken Nancy Love. "Elite" is the word used most often to describe her and her original Women's Auxiliary Ferrying Squadron. The WAFS consisted of around 25 privately trained women who were already ferrying light planes for the Air Transport Command before the first WASP pilots completed Air Corps training.

Jackie Cochran was made chief of all women pilots, while Nancy Love remained boss lady for all girls in the Ferry Command. So both the original WAFS and the Air Force-trained WASPs served under the same umbrella, and all wound up being called Women Airforce Service Pilots. Together, we broke the barrier against women pilots in the military.

The first class of WASP trainees at Houston wasn't exactly a cross-section of American womanhood, because in order to make the program, the first trainees had to possess at least 200 hours of flying time. Accumulating those hours took a lot of money, and our country was still in a depression.

Several heiresses, a French countess, an admiral's daughter, and a Hollywood stunt pilot signed up. Half of them were married. The average age was 26. Twenty-nine women entered the first training class, and 23 of them graduated. Physical facilities were woefully inadequate when that first class entered training in Houston. In fact, they had no place to live! One woman found an apartment over the gatehouse of an estate belonging to family friends, the DuPonts.

Marion Fleishman of the Fleishman shoe fortune arrived with two afghan hounds. She, the French countess, and another blonde heiress each rented a suite in an uptown hotel. They occupied an entire floor.

Some of the young women teamed up and found rooms in a wealthy section of Houston, where patriotic families turned their mansions into boarding houses. Housemothers accepted the girl pilots' canned explanation that they did classified work for the AAF, but other boarders viewed them with frank suspicion. Some trainees had to live in a tourist court for $2 a night and promptly came down with "crabs." If that wasn't bad enough, the courts had limited hot water and few single beds. But the girls soldiered on.

The airport offered a single toilet for 28 women. Some days the girls had only a candy bar and a Coke for breakfast. They lunched at a depressing airport cafe where several girls got food poisoning.

Unpredictable weather was the worst thing about Houston. Rain and fog kept training behind schedule for weeks at a time. So operations moved to a training field in Sweetwater, Texas, a wide-open place with no fog and little rain. Sweetwater was truly the wild and wooly west, "out where the rattlesnake rattles and the buzzard builds its nest."

One trainee was up in the air in her open cockpit trainer when she spied a rattlesnake on her wing and crawling toward the cockpit. She had the presence of mind to put the plane into a sideslip and the rattler went sailing off into the air, spinning like a helicopter blade.

I spent six months in flight training at Sweetwater's Avenger Field. We were the first all-Sweetwater class, and we overlapped the departing class of male cadets, which created interesting situations. Their barracks were alongside ours, with no shades on the windows. We learned to dress and undress while crouched down with bent knees, like Groucho Marx. If we stood up too soon, a chorus of wolf whistles filled the air. Today that would be considered sexual harassment, but back then we thought it was funny.

The C.O. didn't see the humor in all this and had the windows painted black. Living in the one-board-thick barracks with no ventilation was like trying to sleep inside an oven; we had to open our windows to keep from suffocating. We just needed to remember we were being watched every minute.

At Sweetwater we lived in GI mechanics' coveralls made for men. They seemed to come only in two sizes: 42 and 44. We called them zoot suits, and we each received two of them. The smaller girls had to roll up their pant legs and sleeves, and the suit's crotch hung down around their knees. We were not a military looking outfit.

We kept clean by wearing a dirty zoot suit into the shower. We'd arm ourselves with a scrub brush and bar of soap, then work the suit over till the sand and grime and sweat were gone. Then we'd hang the coveralls up and wear the other pair while the first one dried.

The zoot suits were our everyday uniform. We wore our "Sunday uniform" whenever a visiting general came to inspect the troops. Our dress uniforms consisted of tan slacks and a white shirt (purchased at our own expense at the department store in town), topped by an overseas cap. Shoes were rationed, so we wore whatever we had left from civilian life. No two pairs were alike. Again, not very military.

Word quickly got around about females training at Avenger Field. Although the field was closed to outside air traffic, cross-country pilots suffered mysterious power failures over Sweetwater, necessitating emergency landings. At least two movie stars had "forced landings" while I was at Avenger Field: Buddy Rogers and William Holden. They were handsome beyond belief to us in our deprived state. But they were whisked away before many of us on the flight line could get more than a quick, excited glimpse.

Most of our excitement came from the airplanes. Although all of the trainees were licensed pilots before being accepted for Air Corps training, most of us had never flown anything larger than a Cub. So our primary trainer, the open-cockpit low-wing Fairchild monoplane PT-19, looked enormous to us. (Later in the program, they switched primary trainers to the Boeing "Yellow Peril" Stearman biplane.)

We ate off tin trays, slept on lumpy cots, and shared a latrine with 12 other girls, without so much as a shower curtain for privacy. We marched everywhere we went and belted out lusty songs to keep our spirits up, because waiting for us was the intimidating Basic Trainer, and we had to be psychologically ready. The big plane was called the "Vultee Vibrator"—the BT-13 or BT-15 (depending on horsepower).

We flew every day (often seven days a week) and sometimes, in panic-stricken terror, we flew at night. After basic training we went on to Advanced in the all-time favorite trainer, the AT-6, the powerful North American dreamboat, with its classic design and retractable landing gear.

Then, in order to learn to fly multi-engine planes, we suffered through the necessary hours in the "Double-Breasted Cub," the twin-engine UC-78 Cessna, also known as the Bamboo Bomber or the Woodpeckers' Delight because it was made of cloth stretched over a wooden frame.

While most of us didn't like the Bamboo Bomber, we dearly loved the AT-6, which can still be seen today in air shows around the country, frequently re-enacting the attack on Pearl Harbor. The Japanese "Zero" fighter was copied (it is said) from the AT-6.

Once, when I was up in the sky in a practice area in an AT-6 at Sweetwater, dutifully doing prescribed maneuvers solo, a Navy pilot "snuck up" on my wing in his fighter plane (which we called "pursuits" then) and wanted to play. The problem was, his pursuit was so much faster than my advanced trainer that he had to let down his landing gear and flaps to slow down. For my part, I had to give the AT-6 full throttle.

I was happy to cooperate and ready for fun, so we flew along this way, laughing and blowing kisses and playing tag with the clouds. Inevitably, my time ran out, and I had to turn back to Avenger Field, which by now had earned the nickname Cochran's Convent. The name came about when Jackie Cochran found out a few of us were dating instructors on the sly. She cracked down and wouldn't even allow us to drink a Coke with them at the BX (base exchange) any more.

So when that handsome Navy pilot flew away, it was back to the old nunnery for me. Back to lining up, stepping out, march, march, marching everywhere, in our hot, baggy zoot suits, sweltering in ground school and suffocating in the Link trainers (early day flight simulators), eating whatever rationed food was put in front of us in the chow line—including horse meat at least once. (It tasted awful.) We wore out the soles of our shoes as we drilled on the packed, scorched Sweetwater earth and hoped they'd hold up for one more half-soling at the shoe shop in town.

Yet, despite the deprivation, I loved life more than ever, because I was part of something bigger than myself. Our country was at war, and we were doing something to help—something needed, something important, and we were all pulling together. Like most people in the United States during WWII, we WASPs were deeply patriotic and willing to die for our country.

When one of us was killed in a plane crash, more than once we had to take up a collection to send the body home in a plain pine box. A classmate would accompany the casket in a hot, stuffy boxcar in the summer, or shiver through the freezing night in the winter, as she escorted the remains of her fallen sister pilot. Sadly, the bereaved family couldn't drape the coffin with an American flag or even put a gold star in the window, and there was no insurance. All this because we weren't officially members of the service.

In retrospect this seems shocking, but it is well to remember the world was a different place in the forties. Though we felt incredibly fortunate as women to fly Uncle Sam's military planes for the first time in our country's history, we weren't fighting for women's rights; we were fighting for our country.

We were paid a little less than second lieutenants. And sometimes we received a little less respect as well. For example, the Military Police often accused us of impersonating officers, since in those early days we didn't have a uniform of our own. After winning our wings, we wore the same shirt, trousers, wings, and insignia the men wore. So it's no wonder the poor MPs were confused.

Once, when I was ferrying a twin Beechcraft with the top secret Norden bombsight installed, I aroused more-than-usual suspicion among the MPs. I had strapped on a .45 as ordered, though no one showed me how to use it. Thus garbed and armed, I was rudely stopped in an airport and challenged. Usually, the MPs were polite, but this one was sarcastic and disrespectful, even after I produced my I.D. and tried to explain things.

I was exhausted from flying all day with no lunch, dodging thunderstorms, and putting up with an ailing radio, but I tried to patiently explain what a WASP was. (Later, I heard three WASPs were thrown into the brig overnight in Americus, Georgia, for "impersonating an officer.")

I wanted to avoid trouble, especially since I was required to post a guard on the plane at night, which was difficult at best. But, when he leered at me and said mockingly, "Sure, sure, you're a pilot!" my patience ran out. Weary, disheveled, hungry, and in dire need of a rest stop, I pulled myself up to my full height (taller than he was) and barked, "Say SIR when you speak to me."

Startled, he snapped to attention and saluted, saying, "Yes, SIR!" and that was the end of that.

Crazy things happened when we delivered planes for the Ferry Command. Irene, one of my classmates, was ferrying a P-51 on a hot summer day in Texas. It was stifling in the closed cockpit of the Mustang, so Irene decided to take off her shirt since she was alone up there (she thought).

"But even after I took my shirt off, I was sweating," she told me later. "It was so hot I could hardly breathe. So I decided to crack open the hatch.

"It was a stupid thing to do," she admitted, "because I knew how things flew out of the hatch—maps and pencils and things—and ZIP! out went the shirt. It was gone before I could grab it."

She flew along, furious with herself, but luckily she had another shirt in the briefcase under her feet, so she continued along, topless, hoping at least to cool off. All of a sudden Irene had the eerie sensation of being watched. She looked over her shoulder, and sure enough, a grinning Navy pilot flew right off her wing, close enough she could count his freckles.

"A lot of good the clean shirt in my briefcase did me," she said. "I didn't dare move a finger, with his prop about to chew a hole in my wing."

It seemed like hours before he finally peeled off and disappeared. Irene always wondered what kind of stories he took back to his base about half-naked girl pilots. But, despite her embarrassment, Irene still remembered Navy pilots fondly, mainly because they thought WASPs were wonderful, since we weren't competing with *them*.

On the other hand, Army pilots sometimes resented us. Like when my roommates and best friends, Sandy and Shirley, (Shirley was LIFE Magazine's cover girl) were ordered to Dodge City, Kansas, to B-26 school.

The B-26 Martin Marauder medium bomber was a killer airplane before they added extra inches to the wings. It was called the Flying Coffin, and the Widow Maker. It was also nicknamed the Baltimore Prostitute, because it was made in Baltimore and had no visible means of support. So many of these planes crashed in Florida, they had a saying: "One a day in Tampa Bay."

So, with good reason, men balked at flying the B-26. General Arnold decided to shame them into it by training WASPs to fly it. Hence, Shirley and Sandy's orders.

Their flight leader didn't know his new class consisted of females. When he strolled into the Ready Room and saw his new pilot trainees were female, he stomped out in a macho huff. "I'm not flying with any

G.D. women," he snorted.

The girls were insulted and furious, Sandy most of all. "I'll get even with that little pipsqueak if it's the last thing I ever do," she vowed.

And she got even with him. They were married for 65 years.

Sandy and Shirley towed targets in their B-26s out over the Gulf of Mexico for green officer gunnery trainees to shoot at. More than once, they came back with bullet holes in their bomber.

I went as a guest to an 8th Air Force luncheon meeting in Dallas a few years ago, and a man came up to me and said, "I've been waiting for years to meet a WASP so I could apologize for something I did during WWII."

I knew before he said it what his confession was going to be. He'd been a gunnery officer trainee who accidentally put bullet holes in a B-26 a WASP was towing targets with. He still felt guilty about it. I told him if he repented he was forgiven, and I passed his apology along to Shirley and Sandy.

Forty-one WASPs, including Shirley and Sandy, flew the B-26 for more than 9,000 hours without a single accident. Many of those hours were accumulated while flying technical and often dangerous missions.

The experiment with WASPs flying the B-26 was so successful (with men no longer refusing to fly it) that when a similar problem arose with men objecting to flying the B-29, Gen. Arnold and Paul Tibbets decided to call in their secret weapon: the WASPs. The story is told in Gen. Charles Sweeney's book, *War's End*:

> The B-29 was getting a reputation among some pilots as being unreliable and dangerous. [For one thing, it had killed the test pilot at Boeing who first flew it.] A colonel who commanded a group of B-29s in training at Clovis, New Mexico, called Tibbets to ask his advice about the growing hesitancy of his pilots to fly the airplane. Engine fires, in particular, had become commonplace. The psychological effect of the pilots' lack of confidence was feeding on itself.
>
> Tibbets had an idea. He selected two WASPs in our unit and took them in a B-29 to an unused airfield in Anniston, Alabama. It was important that word not get out. He trained them to fly the B-29 in three days. When he was satisfied they were ready,

he sent them on their way with orders to land at Clovis.

Word spread quickly throughout the flying community about what happened next. When the B-29 arrived at Clovis, it taxied to the ramp, parked, and out from the front wheel well stepped two women pilots, smiling and exuding confidence. The message was clear. If women could fly this monster, then men should have no trouble.

True, this was a sexist thing to do, but a remarkably simple way of solving the problem. The colonel at Clovis had no more complaints from his men, and we had two more pilots qualified to fly the B-29.

One of those pilots was named Dora. More than 60 years after she demonstrated that even a five-foot-four inch girl could fly a B-29, she ran into the former base test pilot at Clovis, named Harry, who'd never forgotten her. By then, she was a widow, and he a widower.

So after more than 60 years, Harry and Dora, two old B-29 aviators, were married, and are flying into the sunset together. This was the only time in history a B-29 pilot married another B-29 pilot.

Harry recently recalled one persistent problem with the B-29: sabotage. Oily rags were deliberately left in tight-fitting engine cowlings to start fires. Sugar was sneaked into fuel tanks. Tires were slashed. There were numerous electrical problems with the B-29, especially with defective voltage regulators. Things became so bad the FBI placed agents among the mechanics on the line at Clovis, but no one was ever caught.

WASPs had their share of sabotage, too, including control cables being cut. But we flew on in spite of fatalities, until we were no longer needed. The war was winding down in the spring of 1944, and suddenly the Air Corps had surplus pilots. Without fanfare, the WASPs were disbanded in December of that year.

I'm grateful I had a chance to fly alongside some of the bravest women I have ever known. We loved flying Uncle Sam's planes when he needed us. Now they call the WASPs part of the Greatest Generation. If that's true, then this quote from an unknown source is also true:

"We live in the land of the free only because of the brave, who didn't know they were, until it was thrust upon them."

1943-44 WASP Trainee barracks at Avenger Field, Sweetwater, Texas.

Marion Stegeman Hodgson in front of primary trainer, PT19A at Sweetwater, Texas, 1943.

Marion Stegeman Hodgson Taken from Love Field, 5th Ferrying Group Yearbook, 1943-44.

*Marion Stegeman Hodgson in full uniform.
Photo taken by a sidewalk photographer
at the height of World War II in 1944.
On street of Atlanta, Georgia.*

*WASP Graduation Ceremony: WASPs "At Ease" at
Avenger Field, Sweetwater, Texas, 1943.*

GOLMA

by Ken Larson

Peculiar travel suggestions are dancing lessons from God.
~~Kurt Vonnegut

She was a classic Italian beauty, showing grace and good looks in everything she did. Even in the harsh light of the east African desert, she made my heart race. I called her Golma. On that blazing hot Somalia day in 1986, Golma came to my rescue, just as she rescued many others during the time I knew her. Golma, you see, was an airplane[1] and this story is about her and my African friend who piloted her. Together, they touched many lives along the north coast of Somalia.

I sat in a small grove of trees with three Somali men just outside the coastal town of Boosaaso. Because their staccato English was mixed with Arabic, Italian, and the local Somali dialect, I missed most of what they were saying. Although the temperature hovered around 130 degrees, the shade provided by thin-leaved desert trees felt surprisingly refreshing. The men sitting around our table engaged in lively debate, typical of the Somalis I knew. Each argument was emphatically punctuated with strong hand gestures and an occasional wave of the hardwood sticks these men of the desert carried. The two older men wore white head covers, not quite turbans, yet wrapped and tied so they stayed in place and provided protection from the intense sun. Two wore white kaftans that appeared to be woolen. The younger man wore

---

1   Golma is an Italian-made Partenavia. She has six seats and twin piston engines. Her registry is British and her ID is G-OLMA. The "G" indicates Great Britain.

Western clothing. Two old rifles leaned against the tree trunks to my right. They belonged to the men in white traditional garb. Each man obviously had strong feelings about the topic. They were so adamant; at times I almost expected a fistfight, or to see one of the men grab his rifle and start shooting. But I'd seen these animated discussions before, so I didn't worry.

I was constantly amazed at the physical toughness of these people, though they appeared deceptively thin and frail. We were eating meat cooked over an open fire by two women who labored silently. They were tall and strikingly beautiful, with rich chocolate-colored skin and high cheekbones that would be the envy of any fashion model.[2] Their black eyes sparkled in the harsh sunlight, and they moved with elegant grace. Both wore sheer, brightly colored garments, one blue and white, the other primarily green. Gossamer scarves framed their faces. The dresses and scarves hung in smooth cascades and rustled gently in the desert breeze as the women worked. Slim, firm bodies silhouetted through their garments were evidence of the hard life in this part of the world, making it impossible to guess their ages. People grew old quickly in the desert.

One of the women set a piece of brown Kraft paper with pieces of lean, roasted goat in front of me. A bowl of boiled spaghetti noodles was on the table, and the men took turns scooping out handfuls. Knowing it might be contrary to Muslim custom, I didn't reach for spaghetti because it wasn't placed in front of me. None of us had silverware, and we ate with only the fingers of our right hands, as was the Islamic rule.

I did reach for a nearby mug of milk and was quickly stopped by one of the Somali men. I feared I had violated one of the Islamic rules related to food.

"Camel," he said, pointing to the milk, then toward some animals standing nearby.

He shook his head and wagged his finger. He said something to me I didn't understand and all the men laughed. The two women turned away from me, politely disguising their giggles. I was embarrassed until one man's graphic gestures clearly indicated I would suffer a bout with

---

2   Iman, the super model turned cosmetics mogul, is a Somali. She was one of the first black fashion super models.

diarrhea should I drink the camel milk. Later, I learned unless your body was accustomed to camel milk, drinking it could lead to gastrointestinal disaster.

Instead of the camel milk, I was given a badly rusted metal cup. The woman in green poured a yellow drink I recognized as the drink of American astronauts—Tang. I wondered silently whether drinking Tang made with water from an unknown source and served in a rusted cup was any safer than the camel's milk. I sipped my Tang in the Islamic way, though I wanted to chug it all down.

As I drank, an image of something I'd seen in Djibouti on the edge of the Gran Barre came to mind. A group of nomads with about thirty camels had moved to a river that normally was a dry *whaddi*. Somehow, they knew when water would come to this place, possibly flowing from flash floods in Ethiopia. The camels lounged in the river water, some rolling and rubbing on their sides. Kids stood with and climbed on the camels, apparently washing the beasts. Just downstream from the bathing camels, women filled water containers and carried jugs and water bags to the shore, later to be loaded on the camels. In spite of wondering if my Tang was made with camel bathwater, the drink was refreshing and I escaped any stomach problems.

In the distance stood a mountain range, rough and composed of sheer rock resembling black basalt. To the southwest, sheer cliffs rose from the sand, sentinels standing hundreds of feet tall. I could see down onto the village of Boosaaso. Beyond, to the north, lay the ocean, the Gulf of Aden. Behind me to the east lay desert, sand, and rock as far as one could see. Everywhere, the air close to the ground shimmered and danced in the heat waves. A mirage, appearing as a vast, cool lake, floated between us and the black mountains. I understood how hapless wanderers dying of thirst were fooled by these optical illusions. Eight camels stood about fifty yards downwind (luckily), occasionally snorting. Two of them argued and spit at each other.

No wonder these people were so stoic and tough. They lived in a harsh land. Nothing was soft or forgiving, and even the few sparse plants seemed ready to inflict grievous injury. The ocean looked intimidating, as if the turquoise water might be boiling hot. Yet, these rugged people had kindness beneath their hard exteriors. They were gracious and, without discussion or fanfare, invited me to join them for lunch in their tiny oasis among the trees. Perhaps it was nomadic courtesy to rescue a traveler stranded in a strange desert.

I felt as if I'd interrupted a business meeting, perhaps negotiations to buy and sell the eight noisy camels or a few of the scrawny cattle we often saw along the north coast. I never did understand what they were discussing. And it was difficult at times to tell whether they were discussing terms of a deal or arguing about matters of life and death, because the conversations were so animated and intense. Either way, I enjoyed the scene and felt lucky to be among these people, especially under the circumstances.

I was stranded here on the north coast of Somalia because of pilots and an airplane. As part of my work, I traveled frequently between the *Republic de Djibouti* and Mogadishu, Somalia's capital of about a million people. I was part of a World Bank fishing project, and we operated two ships out of Djibouti, the only suitable port in the region. Since the project was for the benefit of Northern Somalia, our headquarters was in Mogadishu. On this particular trip, I headed back home to Djibouti, traveling on Somali Airlines.

The plane was a high-wing, twin-engine Fokker, flying between Mogadishu and Djibouti, with a stop in Boosaaso. The flight over the rough mountainous spine of Somalia was interesting and the starkness of the land beautiful to see. I sat on the left side, about midway, directly under the wing. As we descended toward final approach, the crew lowered the landing gear, which was long, extending from the high wing with a heavy, stiff strut and large tire. The left main gear was directly to my left, only a few feet beyond my window. The approach was turbulent, as usual, because of unstable air caused by heat rising from the sand and rock below. As we bumped toward the ground, our shadow moved across the land and appeared to climb up to meet us, closer and closer. When the end of the runway passed under us, the pilot raised the nose, adjusted power, and set up for our touch-down as our shadow connected with the left main gear tire. I was fascinated, watching the ground moving up toward the left tire. Closer, closer, then touch-down.

The left tire touched the asphalt, but to my surprise, it didn't start turning; rather, it smoked and skidded. The brake was locked. The smoke increased, then the tire burst into flames. Thick smoke poured from the rubber. Pieces of steel-belted tire flew off, some hitting the bottom of the wing and wing flap, making loud thudding sounds. I couldn't tell if the wing was damaged. Then, the tire disintegrated and vanished. The big Fokker swerved hard left, and the bare metal wheel trailed sparks like a deadly meteor. The wheel was being ground down by friction. We were sliding toward the sand and rock alongside the runway.

A few passengers screamed as they were thrown sideways. Tray tables dropped and overhead compartment doors popped open. Boxes and suitcases tumbled to the floor, creating a sense of chaos and intense danger. I imagined the left main strut going off the pavement, catching in the sand and rock, and causing the Fokker to flip over. Fire would engulf the plane. Death close at hand. Instead, through piloting skill or sheer luck (sometimes it's better to be lucky than skilled), the pilot held the plane on the hard runway and stopped.

In a situation like this, most pilots I know would have stayed put, afraid of causing more damage by moving the plane. Taxiing around with a damaged main gear is a bad idea, not to mention it takes an inordinate amount of power. Numerous nightmare scenarios can become reality, including the gear actually collapsing and the wing hitting the ground, bending and totaling the plane. But, our pilot pushed the throttles forward, taxied the plane off the runway onto a narrow taxiway, and parked beside the small shed that served as the terminal. I was surprised they could move the plane at all with the ground-down wheel, which bumped along as we moved. The pilot must have landed with the left wheel brake locked, which was now free to let the wheel turn and thump.

Typically, on commuter planes, a flight attendant herds the passengers off. Pilots run their shutdown checklists and deplane *after* the passengers. We didn't have a flight attendant, so the ten of us stood, gathered our belongings, and moved toward the front of the plane. One of the pilots opened the door and extended the stairs. I expected him to step back and let us out first, but I was wrong. The co-pilot opened the door, extended the stairs, and both pilots deplaned, leaving us passengers to fend for ourselves. When I finally got down the stairs, the heat astonished me. Mogadishu is just north of the equator and it is hot, but nothing compared to Boosaaso.

I found the pilots seated in the shade under the right wing, positioned as if they didn't want to be within sight of the destroyed left landing gear.

I approached the two Somali pilots.

"Are we going to change the wheel and go on to Djibouti?" I asked.

"No other tire. We'll just stay here in Boosaaso for now," the captain replied.

"Don't you have an office here or a storage room with parts and supplies, or something like that?"

"No."

"When will we get another flight?"

"Don't know," said the co-pilot.

The captain added, "Not today. Maybe not tomorrow, too."

I looked around at the grim sight: sand, rocks, a small town of sand-colored buildings, and certainly no hotel anywhere close. My fellow passengers, all of whom were Somalis, disbursed, as if they'd been swallowed by the sand like characters in a horror movie. I did the only thing that came to mind. I picked up my small bag and walked solemnly toward the town center with no goal in mind. Sometimes the greatest adventures start from such humble beginnings.

A Toyota King Cab pickup (the preferred vehicle of Somalia—next to the camel) stopped beside me. The Somali driver asked if I wanted a ride.

Dancing lessons from God, I thought.

After hopping into the truck bed, we rumbled down the rock-strewn road, bouncing through town and up a gradual incline. The driver stopped not far from town and, after asking permission from the people seated among the trees, led me into the shade where we joined the men who were eating goat.

A few hours later, I saw Golma. I looked up from my goat meat and Tang, and squinted into the cobalt sky at the sound of an airplane. She flew low across town, gracefully banked right, and turned toward the Gulf of Aden. The sun glinted off Golma's right wing as she turned, and I smiled, knowing my ride out of Boosaaso had arrived.

Big Dave van Wyck turned when I walked toward the plane. He'd parked Golma near the shack and was staring at the Fokker's destroyed left wheel when he heard me coming.

"Must have been an exciting landing," he said, the ever-present grin gleaming across his darkly-tanned face. Dave was a huge man who almost had to fold his body into any plane he piloted. He was one of those gentle giants, large enough to be intimidating, yet with a sweet spirit and a kind soul. Dave had served in the Rhodesian army during the guerrilla war and remained in Zimbabwe engaged in any number of ventures, including farming and mining. When prodded and plied with a few beers, Dave could tell wonderful stories about growing up in Africa, elephant migrations, and fighting off big cats. He now had a home on the shore of Lake Kariba.

Big Dave worked for a British company with a U.S. AID contract to conduct inshore fishing studies on the Horn of Africa. They built a camp near Bereeda and Alula at a site called Bolimog, situated right on the point of the horn. Their project camp was built at the base of *Capo Elephante*, a huge elephant-shaped rock formation guarding the south side of the entrance to the Gulf of Aden from the Indian Ocean. Because of the project's remoteness, the company needed reliable transportation and logistics support. Golma fit the bill well. With high wings, the engines were further above the ground, less likely to suck in large quantities of sand, and visibility to the ground was superb for everyone on board. Without passengers, she was able to carry considerable cargo, even in the incredible heat.[3] Golma, therefore, became our lifeline along the north coast. And Big Dave brought her to life, a synergy of man and machine one seldom sees.

"How'd you know where I was?"

"I didn't exactly. Annie heard about the Fokker on the SSB radio, and I came over to see if I might get some spare parts or salvage. We thought it crashed. Your base in Djibouti radioed and said they thought you were on it. They weren't sure, either."

"Are we going to Djibouti?" I asked.

"Just now," Dave replied. His large hand swooped down and snagged my bag as a bald eagle might hook a trout with its huge talons. Dave

---

3 In high temperatures, an airplane is less efficient and requires longer runways to take off and land because the air is less dense than at cooler temperatures. Often, to compensate, the pilot has to limit the load to reduce weight. Djibouti has a long runway, built for the French Air Force Mirage fighters and used by Air France's Boeing 747s.

effortlessly swung my bag over his shoulder and walked toward his plane. I took about two steps for his every one. I considered whether my shoe soles were going to melt into the asphalt on the ramp. Frying eggs on this pavement would not even be a challenge. *Damn, it's hot here,* I thought.

We loaded up and spent the next couple of hours flying over the rough, largely uninhabited north coast of Somalia.

* * * *

Although project support was Golma's primary mission, I had great times with her. Dave and I saw amazing things during our flights across the north coast. One brilliant, sunny day around noon, we'd just passed Berbera on our right. The noonday brought clear visibility down into the Gulf of Aden, allowing us to see deep into the water. I watched the sandy coastline and clear water pass under us.

"Dave! Bank to the left and look down. What are those things?"

Near shore, we saw what appeared to be hundreds of sheets of plywood floating on the water. Dave swung Golma into a descending steep turn, and we circled the site at a low altitude. What seemed at first to be plywood turned out to be hundreds of manta rays, many of them eight or nine feet across. We were close enough to see their wings move gently and their long tails stretched outward, like deadly barbed whips. We had no idea how far below the surface they hovered. Maybe this was where they mated.

In addition to sea life like these rays, smugglers' and merchants' *dhows*, some motoring, others traveling via their unique triangular sails, were common sights across the Gulf of Tadjoura and the Gulf of Aden. Most often, they sailed to and from the Red Sea.

Perhaps the most unique, spectacular sighting occurred the first time I rode with Dave from Djibouti to their camp on the Horn. We had extra fuel, so Dave flew east past Bolimog at an altitude lower than *Capo Elephante*. We were less than 100 feet over the waves as Dave made a wide, arching right turn around the point of the Horn of Africa and out over the Indian Ocean. As we rounded the Horn, turning southward, we sighted a great blue whale, the largest animal on earth. It floated on the surface, perhaps basking in the sun. As we passed overhead, the giant rolled slightly and sounded. When it broke the surface a few minutes later, we were close enough to see the fountain of spray from its blowhole. With all of the stories and TV shows I'd seen about whales, nothing prepared me for the sheer size and spectacular coloring of this monster. Dave kept the whale in view for several minutes. I was in awe and neither of us said a word. As if tired of our interference, the majestic animal disappeared below the surface. As it swam deeper, light refracting through the seawater made the whale appear bright turquoise and shimmering.[4] This was a cosmic view, turned surreal by the play of sunlight through clear, tropical ocean water.

"Magic," I whispered.

Dave added power and made a climbing left turn toward the rocks on the Horn, approaching from the east. We didn't have the distance or the performance to climb over, so he flew us next to the rocks, so close I could see every detail.

"A person could be killed up there. That's damn rough country," I said.

"We'll climb up there tomorrow. Okay?" Dave looked at me. He was seeking a reaction, I guess. We never did make the climb, so I don't know whether Dave was kidding or not.

I imagined climbing those rocks and cliffs in the intense heat, but guessed the view from what seemed like the edge of the earth would be incredible. My thoughts ended when Dave pulled the power to idle, made a few other adjustments, and set us up for a landing. He descended over the water and, at about 500 feet above the glassy cove, turned Golma toward the shore.

"Tray table and seat backs in upright and locked position, mate. We are on final approach to Bolimog International," Dave announced.

I looked ahead, and all I saw was sand backed up by the huge impenetrable rocks of *Capo Elephante*. Just above the waterline, a sharp rise in the sand and rock jutted up about 25 to 50 feet, or more. The rocky beach sloped upward, away from the shore toward the sheer rock cliff beyond.

"On final, but for what runway? I don't see a landing strip." I tried to

---

4   These whales can grow to 110 feet in length, which this one appeared to be. As of a 2002 estimate, fewer than 12,000 Great Blue Whales remain on Earth, and we were fortunate to see one of them.

sound calm.

"Just there. Straight ahead. We built it ourselves. Just packed down the sand and moved tons of rocks. Some were bloody boulders the size of trucks. Problem is, there's no go-around from here if things get dicey. We take off the opposite direction. One way in, one way out, and you hope you aren't too fast or too high to land. Might be able to make a go-around with full power and a hard right turn, but I don't plan to test it. Daytime landings only. Wouldn't dream of coming out here at night. Here we go." Dave focused on flying and verbally ran through his landing checklist. I suspected that, like many pilots, he did the checklists out loud even when he was flying solo.

Golma's main gear touched down smoothly and she rolled forward, slowing as Dave applied bakes. The surface below us was smooth. Dave made a left turn onto a clear area of sand, applied left brake and right throttle, and quickly swung the plane around, creating a tornado of sand behind us. After shutting her down, I helped Dave cover the plane with a lightweight tarp.

"Sand plays hell with the mechanics and the windows," he said. "We do the best we can to protect the plane, but nothing can keep this bloody sand out. It gets into everything. And, I mean everything, if you know what I mean." Dave smiled, tugging at the crotch of his khaki shorts. "Leave the tarp off the wings. We have fuel coming."

A battered truck with a rusted tank on the bed bounced up to the plane and a Somali jumped out, having no door on the driver's side to slow him down. He grinned at Dave and grabbed a large rock, which he used as a chock, quickly placing it to block the wheel from rolling. I was introduced to Ashur, local entrepreneur and owner of the best local water well and the only water delivery truck in the Bolimog/Alula area. Ashur worked part time for the Bolimog Camp and had various enterprises going along the North Coast. Local stories said Ashur's water tank truck had no brakes, so when he made deliveries he followed a specific route, always moving uphill at the right time, allowing him to stop where he needed to. And, he always had the rock to block his truck from rolling once he stopped.

"*Insha Alah*," (If God wills,) he would say in response to the question whether driving a tanker truck without brakes might be hazardous.

Using a hand pump and garden hose, Ashur and Dave topped off the wing tanks, carefully pumping aviation fuel through a three-chambered box with fine-screened filters to catch sand and rust from Ashur's tank.

I'd seen one of these filters before, on the Island of Guam. A sailor named Dick Justice had one he called his Tijuana filter. The device was about two feet long and eighteen inches wide. Fuel poured into the first of three chambers, flowed through the others, and emptied out into the fuel tank. The three chambers served as settling pools for heavier material, such as rust and sand. Fine-meshed filter screens, each finer than the one before it, divided the three chambers to catch debris that could cause engine failure. About a year after I borrowed Dick's filter on Guam to fill my boat tanks with diesel fuel, he was found dead on his sailboat, naked and smiling, adrift on the western Pacific Ocean. I'd never seen him when he wasn't smiling. I shook my head, thinking of him smiling even in death. He knew how to live and how to die.

We finished covering the plane, then walked to the first of three, single-story, temporary buildings. I told Dave the story about Dick Justice and his filter on the way.

"Where's Annie?" Dave asked the only person in the room, a Scotsman named John.

"Went off with soup and medicine about five miles south. Went to see the nomads. She took the Toyota."

John was the master fisherman for their project, which involved inshore fishing and training the Somalis to use small boats and fish for near-shore and bottom species. John didn't look up from the work he was doing in the kitchen area. He was scrubbing and cleaning—and swearing like the Scottish sailor he was. We were in a single, large day room set up to accommodate several people. The room was sparsely furnished, but bright and clean, with a few chairs, two sofas, and an entertainment center holding a TV, VCR, and stereo. Three large windows opened to a spectacular view of the beach and ocean beyond. One corner of the room served as a kitchen, with a large double sink, two chest freezers, a refrigerator, and a cooking stove.

"What the hell are you doing?" Dave asked.

For the first time, John looked up. Incredibly, the front of his shirt and his face were covered with an unidentifiable pink-colored, splotchy substance with a composition similar to spoiled Tapioca pudding. Pieces of the mushy gunk slid from his hair, plopping onto his shoulders and the kitchen counter.

"What the hell does it look like? I'm cleaning the Skill saw and m'self. Gettin' set t' fix you and Annie a nice seafood dinner, fer Chirssakes. Bloody hell!" John went back to cleaning, digging at the saw with a kitchen knife, spitting into the sink from time to time.

After a great deal of prodding and a few insults, Dave got John to confess. He'd taken a large frozen tuna out of the freezer in another building. He decided to cut it into steaks for tonight's dinner, but since it was rock hard and impenetrable to a normal fillet knife, he pulled out the electric circular saw and went to work on it. As the power saw ripped into the frozen fish—flesh, skin, and guts were thrown everywhere. The saw was clogged, the motor packed with mush, and John thoroughly splattered.

Annie was Dave's wife, a registered nurse from the UK with years of experience in Africa, including their rural Zimbabwe home. Annie kept all the leftover food from meals at the camp and made it into huge pots of soup (Annie's soup) which she froze and kept in serving-sized plastic bags. Nomads moving with their camels[5] and cattle through the area had learned about Annie and if any needed medical care, they would send a messenger to the camp and get her to come meet them.

She'd seen all over Africa how nutrition was a major cause of problems for children. These nomads' children were no exception; thus, the soup. She would make all the youngsters in the camel camp eat her soup, then she'd examine them and treat the ones needing help. She told me most of the treatment was for injuries that should be minor, but often resulted in infections for these people because nomads lacked necessary sanitation and lived in such rough conditions. Cigarette burns and infections from face cutting were common. The cigarette burns on the chest and stomach were frequently caused intentionally, as many of these people believed an illness of the chest or stomach could be "burned out" of the body by applying fire. Cigarettes were plentiful and easy for them to use. Face

---

5 Somalis don't ride their camels, as you see people do in Saudi Arabia, unless the person is ill or injured. They are used as pack animals. Nomads still move across the area, ignoring borders and national disputes.

cutting infections resulted from the tribal practice of making a specific pattern or number of small scars on young people's faces to identify their familial connections. Annie, like Dave, touched many lives in that remote part of the world.

Only a few places exist on Earth where you can fly an airplane at an altitude below sea level. Death Valley, California, is well known to Americans, and I've landed there several times near the Salton Sink at an elevation of about -148 feet MSL.[6] Another such place, we discovered, is just outside of Djiboutiville at a geologically-active area called the Ghubbet.

When Big Dave learned of this geological fact, not wanting to miss such an opportunity, he got a few of us together and we squeezed into Golma and took off from the main Djibouti airport. He got an altimeter setting from the French control tower operators to make sure our altimeter would be as accurate as possible for the event. We taxied through simmering heat waves rising off the tarmac toward the main runway. Two *Mirage* fighter jets roared down the pavement and flew low, building up speed before pulling up in steep climbs.

We were heavy with five people in the plane and it was hot — damn hot —as usual. Our takeoff roll was so long, I wondered if we'd get airborne at all. Dave flew low over the coastline, gaining speed. We initially flew in the direction of the Red Sea. Below, we saw several people with camels on shore and a few *dhows* just off the beach. I wondered if they might be smugglers taking Dunhill cigarettes and Johnny Walker to Saudi Arabia, a practice I learned was quite profitable.

When Dave made a steep left turn and headed inland, he also initiated a descent. We all sat in suspense, anticipating the moment we would pass zero on the altimeter, as if we expected something earth-shattering to shock our nerves and twist our minds. We were excited by the unusual experience and our excitement was fed by the energy of five people focused so intently on a single event. We fed on one another's enthusiasm

---

6   MSL means Mean Sea Level and is the height a plane is flying above sea level. Compensation for changes in barometric pressure is done by the pilot using an adjustable barometric setting on a plane's altimeter. AGL refers to Above Ground Level, indicating how far above the ground an airplane is flying. In the Rocky Mountains, for example, a plane could be flying at 12,000 feet MSL, yet only be 100 feet AGL. In parts of Death Valley, one could fly at 100 feet AGL, yet be at -50 feet MSL, or below sea level.

as the altimeter's needles rolled down toward zero. We flew out over the green water of Lake Assal, with an elevation of approximately 157 feet below sea level. And, it happened. The altimeter recorded the moment we flew Golma below sea level. Nothing! No indication whatsoever. No one's earth shattered and none of our minds were twisted by mysterious forces. Nothing happened. Dave and Golma just kept flying.

I recall the first time I went supersonic while piloting a military jet. It was anti-climactic, because absolutely nothing happened there, either. I experienced no shuddering or vibrations like we saw when Chuck Yeager went supersonic in the movie *The Right Stuff*. The only indication inside the cockpit that my T-38 had broken the sound barrier was a dial on the instrument panel showing a speed greater than mach 1.0.[7] The same thing happened with our descent through sea level. The needles on Golma's altimeter met at zero, and then dropped. Nothing else. But, the lack of drama didn't stop us from cheering and talking about it later over beer.

In that part of the world, pilots had few regulations to worry about. Still, to maintain currency and proficiency for the British-registered plane, the pilots had to complete annual check rides and maintain their British pilot certificates. A certified flight examiner was brought in from England to do evaluations on Dave and Peter Mirin, another pilot who occasionally flew Golma. They conducted this check ride in Djibouti airspace. When the examiner arrived at the plane, Dave and Peter were already aboard, with Dave in the back seat and Peter in the pilot's seat. Dave announced it was time to put on headsets. The check pilot obediently put his on, adjusting the boom microphone. When he was satisfied, he looked to the left at Peter. In shock, the check airman turned and looked back at Dave. He started laughing and was laughing so hard he had to get out of the plane so he could stand and catch his breath. Peter and Dave had each put a bra on their heads, with the strap under the chin and the huge cups over the ears, mimicking the earpieces of aviation headsets. I not only drove them to the airport that morning and stood by to see this spectacle—I was at the bar the night before when

---

7   On a few occasions while flying formation and going supersonic, if the sun was just right, I could see the shock wave breaking off of the other plane in the formation. It is the separation of the shock wave that creates the sonic boom for those on the ground to hear.

they *rented* these two bras from an Ethiopian bar girl named Lulu. The bra cups were huge for a reason.

On a serious note, though, Golma, with Dave at the controls, performed important missions. I was indirectly involved with some of them. Our base in Djibouti and the project camp at Bolimog each had a single sideband radio, and we had twice-daily radio conferences. This was before cell phones and the Internet, and we had no telephone links outside our neighborhood. We also had transportation in Djibouti and a driver named Hamud, who always wore white shoes and sported a silver front tooth. I asked Hamud one day where he first learned to drive.

"Jeddah, Saudi Arabia," he answered. "Driving a taxi."

With our white Toyota King Cab and Hamud, we were ready to support Big Dave and the people from Bolimog when they needed something in Djibouti.

During one morning radio conference, John, the Scottish fisherman, said Dave had left early, flying to Berbera to bring an Italian worker to Djibouti for medical attention. We got a call from Dave as he came across the border, asking us to meet him at the airport with the truck. Hamud drove and I rode shotgun. I sometimes caught myself staring at Hamud's silver tooth, trying to imagine the decision-making process behind having it installed. When we arrived, Dave was taxiing up to the temporary hangar they used to keep the plane out of the intense sun and the sand.

Dave spilled out of the cockpit, seeming to double in size as he stood, and stretched his stiff muscles and lower back. He helped a rather small man out of the copilot's seat and held the man's arm as they negotiated the deplaning and walked to our truck. The man was an Italian construction worker employed on a road-building project just south of Berbera, Somalia. He was stocky and strong, dressed in a white tee-shirt and black slacks. He held his left eye closed. We put him in the back seat of the King Cab pickup, and I sat on his left side. As we drove into Djiboutiville from the airport, I got a glimpse of this poor man's eye. He opened it just enough for me to see swollen blood vessels that looked as though they might pop at any second. His entire eye was brilliant crimson, the eyelids thick and blotched with color, the shades of a horrible bruise.

This Italian worker had complained of an itchy and watery eye for a few days, but when the pain became so severe he couldn't work, they notified two Americans who maintained a secret military runway at Berbera.[8] These guys knew Dave and had a way to call him to the rescue.

Djibouti, like much of Africa and the Third World, has two systems of health care—one for the locals and another for the wealthy and the whites. In this case, the whites were the French Foreign Legion. They had their own hospital.

I'd seen the other hospital and found it disgusting—a disgrace to the dignity of humanity. Here's how that visit occurred: One of the Somali crew members on our fishing boats smiled when I first met him, but his smile was for my tee-shirt, not me. It was solid black cotton with a red, yellow, and white image of Mickey Mouse on the front. I gave him the shirt, and I think he wore it every day for a year. After a month or so, people started calling *him* Mickey Mouse. The name stuck, and I can't even remember his real name. Mickey had a ship mate named Abdi who was working at our warehouse on the dock one morning.

This warehouse was a giant structure, sitting back from the pier about fifty meters where our vessels tied up. Purse seine nets, tanks, small boats, and tons of cardboard fish shipping boxes were often moved in and out of this building. Gigantic steel doors spanned the entrance, spreading over 40 feet and reaching about 20 feet in height. The doors were heavy steel, weighing tons.

On this particular day, several men worked at the warehouse, moving a huge pile of fishing net from inside to be loaded on the ship. One of the Norwegian crew was operating a forklift, lifting and pulling parts of the net out so it could be fed over a hydraulic block on the ship and hauled aboard.[9] The forklift driver, most likely drunk or suffering from a hangover, backed into the front edge of one of the huge steel warehouse doors. The steel door bumped upward, making a creaking, rumbling

---

[8] I was told by the Americans running this little base, after expressing my amazement at the length and excellent condition of the lighted asphalt runway, that it was an emergency landing base for B-52s and other large American military planes that for one reason or another couldn't make it into Diego Garcia in the Indian Ocean. I wonder how many runways like this we have around the world.

[9] On one fishing trip, the cook on the vessel got his head caught in the net, and it was pulled into this hydraulic block and he was instantly killed. His head was squashed. The captain had his body placed in the freezer and they continued fishing for four more days.

sound as it slid off of the top roller track. People yelled warnings in several languages (including Norwegian) and the door fell like Jack's giant falling from the beanstalk. Fortunately, Abdi understood at least one of the warnings and ran. Unfortunately, and unlike Jack, Abdi was a slow runner and the door clipped him on his right shoulder and back just as he cleared it. Two more steps and he would have been safe. One step shorter and he would have been killed. I thought he surely was dead.

We got Abdi to the local hospital quickly, but couldn't find anyone there to help. The place had peeling paint and missing window screens. Fly-covered things I didn't even want to think about littered the floor. We found a vacant bed in a well-lit, relatively clean area and helped Abdi remove his shirt and lie on his stomach. He has several bad contusions, including deep cuts contaminated with soil and paint chips. Mickey and two other Somalis went in search of a nurse or doctor. This hospital, I learned, didn't provide food, linen, or medications. These items had to be supplied by the patients' friends or families. All we wanted was treatment for Abdi's wounds and maybe x-rays of his shoulder, which had begun to swell and turn a mean reddish-blue color. An hour after we helped Abdi onto the bed, a nurse entered with a pan of water and soap and scrubbed the wounds. She scrubbed our friend's battered body as if he were a grimy linoleum floor. Abdi was silent, but it must have hurt like hell. Abdi recovered eventually, but not because of that hospital.

The French hospital where we took the Italian worker, in contrast, was modern, clean, and well equipped. Yes, they supplied food and medicine. After only a few minutes in a clean, well-lighted waiting room, a French doctor led us into an exam room. The doctor spoke no Italian, the construction worker spoke no French, and Dave and I were out of the loop altogether.

I couldn't see the actual procedure because Dave and the doctor blocked my line of sight. I didn't fully understand the diagnosis on the man's eye at first, but I suspected it was serious. The doctor applied several drops to the man's left eye to deaden it and dilate the pupil, I guessed. He used a tool to pull the Italian's eyelids apart, exposing the horrifically swollen eye. I'm not a squeamish person, but even I couldn't watch this procedure. It must have been bad for the Italian, because he sent forth a string of heated words, insulting the doctor, his mother, and the entire nation of France.

A male nurse who equaled Dave's stature, joined us and helped steady the angry patient. This nurse had more tattoos than the cast of a porno film. Even in hospital clothes, he looked like he could handle a shoulder-fired rocket launcher or a tank, or maybe hand-to-hand combat with Rambo as easily as a bed pan or blood-pressure cuff.

In this modern, well-run hospital, the French doctor had plenty of desert experience. He quickly understood the Italian's problem: a piece of the desert had, over a period of days, worked its way around the side of his eyeball and lodged there. The grit of sand had penetrated the man's eye and a bad infection resulted. The pain must have been horrible. I couldn't imagine what was going through the construction worker's mind. What must he be seeing as the doc worked close in with those stainless steel tools?

After the procedure and an hour for the staff to monitor our patient, we took the man to a local bar where he promptly got plastered on Scotch and pain pills. At one point, the Italian tried to administer eye drops. He raised the padded pirate's eye patch the hospital had given him and looked toward the ceiling, dramatically holding the medicine dropper at arm's length above his head. He squirted eye medicine all over his forehead.

The nurse came into the bar. I learned he was an American. Had he been in the doctor's treatment room at the start, we'd have had an interpreter. I asked the nurse why he joined the French Foreign Legion instead of a U.S. military outfit—say, the United States Marine Corps. He said he failed an exam with the Marines. No *Semper Fi* for him. No Marine Corps birthday parties. He ran away and joined the French Foreign Legion.

"Which exam?" I asked.

"Psychological exam," he said.

His eyes drilled holes in me. A shiver ran down my spine.

We watched Big Dave and Golma help many people along the north coast, but one event stands out from all others because of the personal risk and display of Dave's piloting skills.

An American research vessel frequented the Gulf of Aden. I thought it was probably a spy vessel, but conversations with some of the crew made the vessel sound more like a huge party boat. A director at the World Bank told me this vessel, like our project, was America's presence on the Horn of Africa—whatever that meant.

Dave and his plane were in Djibouti on a Thursday afternoon, which in Djibouti was the worst time to accomplish anything constructive. Friday is the Sabbath, when work is forbidden. So, on Thursday everyone would kick back and get high with Djibouti's national drug—*khat*. Some call it the poor man's cocaine.

A load of the drug arrived from Ethiopia by train just after noon on Thursdays. It was quickly distributed around the city and sold in small bunches from tables set up on street corners. Combining the chewed Khat leaves with Coca-Cola enhanced the stimulating effects and allowed people to stay up all night chewing, arguing, and eating.

On this Thursday afternoon, Golma was tucked into her hangar space, and Dave was tucked into a sidewalk cafe having dinner when I got a call from Annie on the SSB radio. An accident had occurred.

Since we had no way to telephone, I had Hamud drive us across Djiboutiville, passing the khat tables and numerous groups of people chewing and drinking Cokes. We found Dave.

"A woman on the ship has been injured," I told him. "It's serious, Dave. Annie said they brought the woman to your place on a shore boat and stretcher. Annie said she did her best, but the woman needs hospitalization right away. It's life and death."

"Bloody hell!" Dave grumbled at his dinner. We were at a Yemeni restaurant that specialized in blackened fish. Dave was thinking, planning what to do. He slapped a few bills onto the table and stood. "Let's go use your radio. I'll talk with Annie a bit, then I'll fly over there. About two and a half hours. Gonna be dark, though. Damn."

After the radio call, I dropped Dave at the airport and helped push his bird out of her nest. Neither Dave nor I spoke. He'd hatched a plan that bordered on lunacy, but it seemed the only available option.

"Tanks are topped off. I'll meet you back here in about six hours, mate. Seven tops. Have an ambulance with you." (I couldn't go with him because he needed space for Annie and the patient on the return leg.) Dave's smile shone in the fading light of early evening. He gave me a thumbs-up from the pilot's window and started Golma's engines.

I wondered if this was a fool's errand, risking Dave's life to save an unknown marine researcher (or marine partier, depending on which stories one believed). Upon his return, Dave told what happened.

Hours later, Dave was descending, staying over the water so he had a reference for altitude, knowing no mountains would reach up and knock him out of the sky. At night over uninhabited land, he could see no lights or other references. And, as luck would have it, this was a moonless night.

If you look at a detailed map of the north coast of Somalia, you'll see numerous towns dotting the coastline from Djibouti to the Horn. Most of these are uninhabited or don't have electricity. Some are ancient villages dating back centuries to the time when the sultans of Zanzibar roamed this region, now reclaimed by sand and salt. For Dave's purposes that night, the coast was not only isolated from the modern world, but was also unpopulated and void of light or aids to navigation.

Optical illusions are a common and are serious hazards when flying in remote areas, because normal visual references aren't available. Hundreds of fatal crashes have resulted from these illusions. Your eyes and mind play tricks, in some cases making you think you're too low, resulting in an unnecessary climb and overshooting your runway. In other cases, especially where the terrain is pitch black but a few lights are present, a pilot has the illusion of being further above the ground than he or she really is. This condition is the most dangerous because pilots have a tendency to descend and fly into the ground, short of the intended landing zone. Dave experienced both of these deadly illusions that night.

During the radio call from Djibouti, Dave had worked out a plan with Annie and John at his camp—a plan involving the ship and her crew. From about seventy miles out, Dave spotted the ship. As he'd instructed, every light on the boat, including powerful spotlights, was illuminated. Since nothing else along the entire coast could make so much light, Dave had created a navigational aid. He steered for the lights, trying not to stare for fear of ruining his night vision or getting vertigo. He constantly cross-checked his attitude instruments and altimeter because he had no real horizon to guide him. Other than the ship's lights, nothing else was visible. He couldn't see the shoreline, or even catch a hint of the rocky cliffs and the black mountains to his right. Dave stayed focused, rehearsing how he would make his approach and land on the homemade sand landing strip at Bolimog.

As he approached, Dave called the ship as planned on his VHF radio, and they doused most of their lights. Not only was it too distracting to leave them on, but he needed relative darkness to take advantage of his ground crew. He called Annie on the radio.

"John and Ashur are ready, Dave," she said. "Sure you want to do this?"

"Coming in just now, dearie. Light 'em up."

Three minutes later, a bright spotlight showed from the back of the project house up onto the rocks of *Capo Elephante.*

"Just so we know where the bastard rocks are," Dave whispered to Golma.

One by one, other lights came into view. Dave took a left turn out over the water, circling the ship which lay at anchor about 400 meters off shore. As he turned back toward the coastline, Dave could actually make out the parameters of his landing strip. At the far end sat their John Deere tractor with its light on bright. Halfway down the strip on left and right sides stood their Toyota truck and Ashur's tank truck. These two vehicles faced the center of the runway, hopefully giving Dave a midway aiming point and a bit of vertical awareness. On the left and right sides of the closest point on the approach end, John and Ashur had set oil drums on fire. This wasn't their first choice, but they didn't have any other vehicles with bright enough lights. Dave worried thick smoke from a burning drum would blow across the runway and obscure his vision. This was a factor they hadn't considered. Now, it was too late to worry. Fortunately, the air remained calm.

Dave saw two small lights, unsteady, flashing upward, then making a little sparkle in the water short of the runway. He knew these were hand-held lights John and Ashur were carrying out into the water. The two men were wading, stumbling through the coral and along the rocky bottom. Dave made another left turn, his final circle before taking her in for the weirdest landing of his life.

Dave wiggled his fingers, forcing himself to loosen his grip on the plane's control yoke. Too tight of a grip, the white-knuckle kind, on the controls causes a pilot to be rough and unsteady in controlling the plane. He ran through his before-landing checks and reviewed how he would fly this approach. He briefly considered his options if this didn't go perfectly.

If he landed long, he would crash into the John Deere, a solid piece of steel. If he went high and couldn't land, making a go-around was tantamount to suicide with *Capo Elephante* standing there, a giant beast formed of solid stone. If he landed too far to the left, he would probably die in a fireball caused by hitting Ashur's fuel tanker. On the other side, the Toyota would be a bad mess, but not as bad as smashing into Ashur's old truck. If he wasn't centered on short final, his fate would involve crashing into a flaming oil drum. Dave forced that scenario from his mind. The plane's high wings brought a bit of comfort, because he didn't think a wing tip would be low enough to hit an oil drum, even if he veered off the runway's centerline.

Dave circled the ship one last time and rolled out on final approach. He now had two lead-in lights that also served as range lights, giving him better directional cues to line up correctly. And perhaps best of all, he could make out the rise of land as it sloped up from the water's edge. John and Ashur were waist deep in water, as closely lined up with the center of the runway as they could get. Each held a flashlight pointed toward Dave. John, who was nearer the shore, also had a bright hand-held spotlight. He focused the beam on the sand and rocks just above the water.

Dave lined up on the flashlights and checked to make sure he had full flaps to help keep his airspeed slow. He would go in at minimum approach speed. He wiggled his fingers again, loosening his grip on the control yoke and the two throttles. The props were set for landing and he slowed, making slight adjustments in pitch and bank to stay aligned with John and Ashur while fighting the mental image of hitting them if he landed way short. The air was stable and smooth over the water, but Dave wondered if the sand and rock would still be hot enough to create updrafts. If so, he would need to compensate or end up drilling his plane into the grill of their green tractor. If he over-compensated and was low, he would hit the edge of land where it rose from the water.

Dave sped past Ashur, not looking at the man, but thankful he was there. Dave kept his focus on the runway, watching his altitude, checking airspeed, adjusting his pitch. He was past John and now could see the rising sand and rock.

It felt like he was too low. Dave was certain Golma would plow her nose into the rocks short of the level landing strip. He pulled back ever so slightly on the controls, fighting the urge to climb quickly. He recognized it was an illusion, caused by flying from total darkness into relative brightness and accentuated by the land rising toward the runway. The illusion made it seem the plane was descending too fast, because the land was moving up toward him.

Airmanship born of experience in unusual flying situations took control, and Dave fought the urge to climb. He steadied Golma as a jockey might settle a thoroughbred before making a break for a strong finish. He talked to her as they moved closer to the sand, reassuring Golma she was going to make it, assuring her he would not be the cause her demise. Together, man and machine, a synergy of effort, flew closer to flaming oil drums and trucks and a solid American tractor.

He was now over the rise, and the land leveled off, creating the opposite feeling of being higher than he should be. Dave fought the fatal inclination to dive. He approached the burning oil drums, thankful the wind was blowing lightly onshore, if at all, even though it created a slight tailwind. His view forward was clean. With his right hand, Dave lifted the two mixture control levers and pulled firmly aft. He had landing lights, but they were of minimal use, so he turned off all electrical power. The landing lights went dark. Both engines died. If he did crash, shutting off fuel and spark reduced the chance of fire and engine damage. With no power, he and Golma had no choice but to land.

Golma handled perfectly, moving exactly as Dave's huge hands asked. She was stable when needed, yet responsive and nimble when called upon to act. And she was strong as they hit the packed sand hard. She made a slight bounce, then settled nicely and responded as Dave pressed the brakes. She came up short of the John Deere.

Dave dropped his hands in his lap and rolled his head back. "Bloody hell. We did it."

We met Dave at the Djibouti airport early that morning, well before sunrise. He was exhausted, but came with us to the French hospital. The French doctors were able to save the woman, but said she'd arrived with little time to spare. She was evacuated on a U.S. embassy support flight a few days later, lying on a jump seat along the side of the cargo deck of an air force C-141.

Soon afterward, we began packing and closing down our projects. Northern Somalia was building toward all-out civil war, with rebels gaining strength and the local authorities becoming more wary of us and making demands we couldn't meet.

After armed men raided the Bolimog Camp, holding Dave at gunpoint and firing rounds into the wall above John's and Annie's heads, they called it quits. John and Ashur loaded about fifty gallons of diesel fuel into one of John's 12-meter fishing boats and headed westward down the Gulf of Aden toward Djibouti, like Huck Finn and Tom Sawyer seeking adventure. They had no navigation equipment and few rations, but three days later, they recognized the entrance to the Red Sea and turned left, pulling into Djibouti harbor on the fourth day. Reports say John bought a case of beer and locked himself in a room at the Sheraton Hotel. We don't know where Ashur went, but I suspect he's a wealthy businessman, probably living in Europe.

Golma's last rescue flight[10] in northern Somalia took Dave and Annie out of the isolation of Bolimog Camp so they could return to their home on the banks of Lake Kariba, back to their lives of farming, mining, and whatever other ventures came their way.

Two years later, Dave was hauling farm equipment with a truck and trailer. Annie and their son Christopher were following in another vehicle driving toward home from Zambia. Dave's truck was attacked by bandits who stole the truck, the farm equipment, and did to Dave what years of flying and night landings among burning oil drums and green tractors could not do. They murdered Big Dave Van Wyck.

---

10 I did a registry search for G-OLMA on the Internet and found that she is still in service as of mid-2006 when she was getting a retrofit upgrade on her engines.

Bossaaso, Somalia. Two young men and their camel on the beach.

Street scene in Mogadishu, Somalia, near beach.

A TWO-FER DAY

by Kevin Kasberg

Abu's nerves calmed as he sat in his yellow Mercedes-Benz truck listening to the idling engine. He felt nervous, excited, as the greatest moment in his life approached. Abu fought against the idea of fear; rather, he was certain the feelings came from pride at being selected for this great honor.

Just after 6 a.m. on this October, 1983, morning in Beirut, Abu shifted his laden truck into gear and drove slowly through the international airport's parking lot. He had an important delivery to make and—*inshallah*—nothing would stop him. Two circuits through the parking lot, then he steered hard and gunned the powerful truck engine. Black exhaust spewed from the stack; the noise of the engine, the gears, and big tires grew louder as his speed increased. Abu's eyes were huge and his mouth hung open wide as he spat out his hatred of America. He fixed his eyes squarely on the concrete-block building beyond the fence and the two US Marines standing guard.

His people had successfully bombed the American Embassy in Beirut the previous April. That was an important victory for Islamic Jihad, but would pale in comparison to what he was about to do. Abu would be martyred beyond any others. He accelerated faster as he approached the barbed wire fence. He could see the faces of the two Marine sentries at the small guard house. They lifted their weapons.

When Abu's yellow truck ripped through the barbed wire and bumped past the guards, he smiled at them. "Not yet," he said aloud, knowing the gas-enhanced explosives in the cargo box of his truck must have shifted. His cargo was the equivalent of 12,000 pounds of TNT, including canisters of propane, butane, and acetylene, which would be triggered by the central explosive device. Abu had little experience with bomb-making, but even he knew this bomb wasn't sophisticated. It was just huge.

"*Allah-u-akbar*," Abu screamed. The Mercedes slammed into the concrete wall, breaking into the lobby of the Marine barracks. Abu detonated his load, lifting the entire concrete-block building off the ground, blasting a huge crater into the Beirut soil, and sending him to paradise. Abu was martyred. Two-hundred forty-one Americans were killed, and untold other people were injured, including innocent Lebanese.

About two minutes later, another Islamic Jihadist rolled a bomb-laden truck down the ramp to the underground parking garage at the French military barracks. The resultant explosion leveled the eight-story building, killing 58 French soldiers and injuring 15 others.

* * * *

Early on a beautiful Mediterranean morning, I stepped onto the deck of my ship. The scene on the USS *Milwaukee* appeared chaotic, but was actually a choreographed ballet made possible by extensive training and personal discipline. Sailors darted around the deck, securing cargo nets on pallets of supplies and equipment. The deck, barely large enough to hold the cargo brought on board, was encircled by pallets shrouded in heavy nets, each with a lifting cable attached. Sailors stacked the pallets in alleyways, leaving only an area in the center large enough for a helicopter. The only clear space was a narrow path from the two on-deck hangars to the small platform for launching and landing helicopters.

The *Milwaukee* was a combination cargo supply ship and oiler, providing a critical support mission to this multinational task force. Each ship within the task force needed supplies of food, fuel, equipment, and other important items. The huge aircraft carriers needed fuel for their engines, as well as aviation and jet fuel for aircraft. Support ships like ours made naval task force operations possible.

The rumble of the engines deep inside our ship was drowned out by sailors barking orders and securing the cargo, a task that took all night. The ship gently rolled in the eastern Mediterranean swells, with occasional sprays of seawater coming on deck. The stimulating sounds and smells added to the excitement I always felt before an Underway Replenishment. The odor of aviation fuel, machinery exhaust, and sea air was occasionally overridden by the smell of sweat when I passed close to a sailor struggling with a heavy load. No one complained.

As I walked toward the waiting helicopters in the pre-dawn morning, the lights of other ships in our huge battle group stretched to the horizon. About a mile behind us and closing fast was the USS *John F. Kennedy*, a 1,052-foot long, 130-foot-wide aircraft carrier. JFK was big; her flight deck sat 80 feet above the water, about the height of a ten-story building, and she carried a crew of 5,000 people and more than 80 aircraft. *Kennedy* was a conventionally powered ship with steam-turbine engines that needed refueling on a regular schedule. The aircraft carrier *America* also was out there in the dark. I smiled, thinking about the amount of power we represented. "Incredible," I said to the night.

French and Italian ships and crews were part of this multinational task force, including the French aircraft carrier *Foch*. This impressive undertaking was put in place by President Reagan in response to the deteriorating situation in Lebanon. The whole damn country was spiraling into chaos and violence. Our embassy had been bombed and suicide bombers killed 300 French and Americans.

The Marine Amphibious Group, with USS *Iwo Jima* as the flagship, was on station near the Lebanese shoreline. The USS *New Jersey,* a re-commissioned WWII era battleship, with its overwhelming 16-inch guns, was there as well.

I knew why we were there—why we constantly prepared for action and maintained our alert status. Just two months before this calm morning, I was on the ready duty flight crew, on standby alert status while the ship prepared to enter the port of Haifa, Israel for a working liberty port call. The calm of that October morning was shattered when the helicopter crew received urgent orders to launch. Having no idea what had happened, I was airborne within twenty minutes, heading at top speed toward the French carrier *Foch*.

As people often do, we speculated on what this was all about, guessing it might be an injured sailor or perhaps a call to transport a high-ranking officer to a meeting. Our orders told us to reach the French ship and work with them—no further details.

We set down on the aircraft carrier's deck with the sun low, but above the horizon. The ship was awash in activity. Planes were launching and sailors hustled around the deck, performing their jobs.

"Something big happened," I told my copilot.

A team of medical personnel, fully equipped with medical bags and equipment met us. Dressed in full battle gear, they started loading with a sense of urgency.

"You must go now," an officer told us in broken English. "This is most urgent. He will tell you where to fly and land." The officer pointed to a young medic who was already inside our cabin.

The medics spoke among themselves in French, so I had no idea what they were saying. I suspected they were reviewing their equipment and making sure everyone needed was onboard. We were airborne within minutes.

"Here." The young medic handed me a map and hastily written orders. We were to fly directly to the *Ramlet al Baida* area of West Beirut, land near the French military barracks at Sidon road, and leave all of the passengers there. We would then return to the *Foch*.

I pushed our Sea Knight helicopter for top speed, heading toward the beach. *Ramlet al Baida* translates to White Sand in English. This was a residential area located on a peninsula pointing toward the northwest. Below us, a gorgeous white sand beach curved along the western side of the peninsula.

"No wonder they call Beirut the Paris of the Middle East. What a fantastic beach," said our crewman from the cabin. He was standing in his usual position, looking out of the open top half of the cabin door.

I flew in toward a spot a few blocks from a small plume of smoke. Though we didn't know what had happened, it felt like something big. Light-colored sand hung in the air and smoke drifted upward from a giant pile of rubble up the street. Debris had scattered for hundreds of meters in all directions around the crater. Later, we learned an eight-story building, housing the French military barracks, had been bombed and totally collapsed.

"There. Put us down there," the young medic said in his thick French accent. He pointed to a vacant lot near the bombsite.

I made a hover landing and within seconds the French medics were off and setting up their triage station. Some were opening cases and setting up equipment before we lifted off for a return to the French ship.

I flew for a solid 12 hours that terrible day, stopping only to refuel. I don't even remember eating. We made countless trips to all of the ships of the amphibious group and the cadre of picket ships and gunfire support ships carrying medical personnel and equipment to shore. We occasionally brought out wounded men.

Indeed, I had a dog in this fight! The Islamic Jihad had attacked and killed my comrades, making things personal.

Now, only two months after that tragedy, even though the sun wasn't up yet, we sat in the cockpit of our Sea Knight transport helicopter. The flight deck was awash in red light as we worked through our preflight duties. Our two Sea Knight helicopters were medium-sized cargo aircraft, able to carry 25 people or a few tons of cargo. Two jet engines powered a pair of large rotors, one at each end of the 45-foot fuselage, spinning in opposite directions. We needed both engines at our heavy weight to climb vertically, hover, or descend vertically. If we lost one engine, our best hope was to fly forward fast enough to control our descent rate.

Ken, my co-pilot, and I would be flying *HW13*, working with another crew in *HW08*. We were ready to move the helicopter into position. Crewmen Petty Officer Ross and Airman Sanchez were in the back to load cargo and operate the cargo hook for external loads. These two men worked from the helicopter's cabin.

The rest of the early crew reassembled in the hangar. We got permission from the bridge to open the hangar door and move the chopper onto the flight deck. I rode the brakes in the cockpit while the rest of the crew pushed. The long rotor blades, usually folded to save space and allow the craft into the small hangar, were now unfolded and fixed in place. We were ready to start engines and my excitement increased. This operation, transferring tons of material to JFK, would require precision, concentration, and unerring teamwork. It also entailed a certain risk. We would make several hundred round trips from deck to deck before the morning ended.

The *Kennedy* moved alongside, still traveling at about 20 knots.

The wind was light this morning, so the breeze we felt came from our movement through the ocean and was from the bow. As air moved over the superstructure of the ship and flowed around the cranes and towers on deck, it created eddies and unstable air, making cargo pickups even more difficult. We always welcomed a good crosswind for this kind of a vertical replenishment (VERTREP) evolution.

The sun peeked above the horizon and the scene that appeared surreal in the dim light became clear and very real. Time to fly.

With our experienced crew, pre-flight preparation took only a few minutes, and we were cleared to start engines and engage the rotors. The jet exhaust smelled sweet, the sound of the turbines whining above us was exciting, and the movement of the long rotors coming to life exhilarated us. Ken and I looked at one another and smiled. Thumbs up from Ken, and we lifted out of our nook in the middle of the pallets. The sky in the east was bright and mostly clear, with a few fluffy clouds. The sight of JFK looming on our left side was incredible. I had to force myself to keep from staring in awe.

The second helo would launch from *Milwaukee* in a few minutes. Not only did we have to coordinate our two helicopters, the air boss from the carrier and the Officer in Charge on our ship coordinated operations of the two ships and their crews. We all communicated on a common radio frequency.

Kennedy was told to move into station a couple hundred yards astern of our ship. Our second helicopter launched, but we would wait to begin our first cargo lift until the two huge ships were in position, connected by lines and fuel hoses. With the two ships connected together (called a connected replenishment) and only about 60 feet apart, fuel hoses and cargo rigs were run between them. All of this was done while the ships moved forward.

I had the left seat, which meant I would be making the drops on *Kennedy*, while Ken, in the right seat, would make the pick ups off the *Milwaukee*. This operation required precise coordination between the two pilots.

We would position our helicopter behind the delivery ship and fly an approach up the wake to the flight deck. The pilot in the right seat, Ken, would fly this part of the cycle, making the pick approach. As he neared the deck edge he would make a left pedal turn and stop with the aircraft perpendicular to the ship facing to the port side, facing the *Kennedy*. During this phase, Ken would fly sideways into the wind, yet holding the pickup hook over the cargo pallet to be lifted. This is tricky work. The hook would be in position over the load to be connected for delivery.

Adding to the difficulty of this exercise, the hook is 20 feet behind the pilot's shoulders and he can't see it. One of the two air crewmen lying on his belly in the cabin and looking through an open hatch in the floor (the "hell hole") manually operates the hook and directs the pilot as necessary. When the load is connected and the hook-up man on the ship is clear, the helo lifts the load vertically until it's clear of other loads on deck. The helo then slides to the aft of the ship in a climbing right turn, flying backward. Maintaining flight with a heavy, swinging cargo hanging below isn't easy under any conditions. Constant attention to shifting weight and overloading the rotors and transmissions was essential and physically draining. We were lucky to have decent weather.

As the helo's head comes through a heading roughly parallel with the ship's heading, the control switches to the left pilot, who continues the climb to the taller ship's deck. Drops on the carrier's deck are made from a position perpendicular to the ship's heading, toward starboard. Again, the air crewman in the hell hole directs the pilot to gently set the load down. Then, he manually releases the load and gives the "all clear." The pilot in the left seat will then climb slightly and commence a sliding left turn to a position roughly between the aft ends of the ships and aligned with the direction of motion.

The pilots switch controls and the right seat pilot approaches for another pick. In a large replenishment operation like this day, two helicopters fly the same pattern. Careful orchestration is critical to avoid hitting each other and to keep the operation moving without a hitch.

If the risk to the helicopters is serious, the danger to the ships is nearly incalculable. Two huge steel ships are connected together, moving forward through the ocean at about 15 to 20 knots. They are simultaneously pitching, rolling, and yawing through the water. Waves strike their bows and the bow waves formed by each ship crash together between the ships' hulls, smashing and banging and sounding like Hawaii's North Shore surf. Sailors are at the helms of each ship, manually steering in perfect formation. Even a small error would mean disaster. A steering error would cause the ships to collide. Move apart too far and the fuel hoses will separate, creating a tremendous fire hazard. Move too close together and the hulls can be damaged, even causing one ship to sink. Speeds have to be exactly matched.

Imagine two 18-wheelers moving down a freeway only inches apart. One semi-truck is a gasoline tanker and the other is the largest tractor-trailer rig in the world. The tanker truck is connected with the big diesel and a hose is connected between them while fuel is pumped from one to the other. That's similar to the connected replenishment at sea.

In 1976, USS *Kennedy* experienced a collision while conducting a nighttime underway replenishment. She was 100 miles north of Scotland, connected with the USS *Bordelon*. The smaller ship lost control and collided with the *Kennedy*. The resulting damage was so severe that *Bordelon* limped back to port, was found to be irreparable, and sold for scrap iron in 1977. The risks were real.

We got into a cadence with the other helo, *HW08,* one offloading on the *Kennedy*, while the other picked up cargo from the *Milwaukee*. We were an aerial ballet and had a steady routine going, breaking only to fuel the helos. Gradually, the rate at which the deck crew on *Milwaukee* could stage loads began slowing, because loads were moved from deeper within the ship. Also, we were able to deliver loads to the *Kennedy* faster than their crews could clear them from the fantail drop zone and break them down. The load rigging equipment, which we called retro, had to be returned to *Milwaukee* so they could reuse it and keep the process going. We had inundated the deck crews on *Kennedy* and the intervals of returning retro were becoming a limiting factor.

Kennedy crews, in order to keep pace, did a good job of keeping a spot open for drops by moving loads all over their aft flight deck. Most of the *Kennedy's* airplanes were on the forward part of the flight deck. A spot on the *Kennedy's* flight deck remained open for us to refuel.

At around 10:30, the pace of the resupply flights had slowed considerably. We now needed to return our nets and other cargo back from the carrier to our supply ship. I brought my aircraft in for a quick retro pick up on *Kennedy*. Our load was a huge bundle of our cargo nets which were hung from longer-than-normal cables. Usually, the length shouldn't be more than two cable lengths, or about 32 feet.

On this hook, Petty Officer Ross was in the hole.

Ross called, "We have hook-up. Hook-up man is clear. Hook is locked. Looks like a big load of pendants. Clear to lift …Up …Up … Straight up … Keep it steady, we have two lengths and more … Up … Up… Three lengths … Up … Fourth length looks like the end."

Ken interjected at some point, "I hate when they bundle pendants end to end like this. Makes control so much harder with all that shifting weight swinging under us. They're just tryin' to save time."

Ross then completed, "The load is clear to go."

We hovered about 70 feet above the carrier's flight deck, putting us in a 150-foot hover above the ocean as we transitioned aft of the ship in a left pedal turn. Just as I was coming parallel to the ships with the nose slightly down to maintain position, the noise level in the cockpit changed—it grew quieter and the rhythm of the rotor beats slowed.

I glanced at the instruments. The triple tach, which indicates rotor speed and engine condition, was decaying. One of our engines was dying. My left hand instinctively reduced collective to keep the rotors' speed from slowing. I commanded, "Ross, pickle the load!" wanting him to release the cargo hook and get rid of the weight.

Ken called, "Number one engine failure."

Ross reported, "Load clear."

In one well-drilled reaction, Ross released the hook, rolled away, and closed the hell-hole door.

We heard the thunk of the hell-hole door slamming closed and Ross added, "The hole is secure."

Lying right beneath the engines, Ross didn't need to wait for a call from the cockpit to know what was happening. He heard an engine winding down and instinctively pickled the load and secured the cabin.

"I've still got it," I told Ken as I stopped the turn and dipped the nose about 10 degrees, hoping to gain airspeed. Being 150 feet above the water became our friend because we were high enough to trade height for speed. We could move forward and develop some lift by descending. With the load gone, we were pretty light. Ken was calling turns (rotor speed) and altitudes while I rolled slightly into a left turn to get headed into the wind and milk the collective, trying to maintain as much altitude and control as I could, hoping to keep the huge rotors spinning fast.

Ken was counting, "Sixty feet," while I was finally getting an airspeed indication.

Our airspeed was building through 50 knots. Over the radio, I announced "*HW13* lost an engine."

Ken was the first to say it, but I observed the same thing. "The left engine is still running."

Indeed it was, but only at idle power. By now, we had both airspeed and enough altitude to reach our ship and an emergency landing. I was relieved we wouldn't be ditching into the ocean. Splashing down into the water can ruin your whole day.

Then, without any command from us, the bad engine came back up to full power, which rapidly increased the rotor speed. I increased collective pitch and started a climb to keep the rotors from overspeeding. After a few seconds, the engine again dropped toward an idle setting, forcing another decrease in collective to keep the turns up.

Ken calmly said, "We can't keep this up." He brought the power lever on our bad engine to idle. We left that engine running.

As good fortune would have it, the ships were now moving away. The lines were clear and *Kennedy* was widening the gap and in a hard left turn away from *Milwaukee*. Tower on *Kennedy* asked, "*HW13*, do you need any assistance? We presently have a red deck, but can have spot three cleared for you in about 15 minutes."

Ken responded, "Roger, we'll let you know."

Then Jim, who was in the tower on *Milwaukee*, announced they had a clear deck and could maneuver to provide about 20 knots of wind on the starboard bow.

Ken and I had a brief discussion about our options. We were light on fuel already and opted not to dump any. Spot three on the carrier wasn't much better, if at all, to the deck on *Milwaukee*, and we certainly didn't want to wait 15 minutes to land. If *Milwaukee* could get us good winds, we would land there.

When we asked, Jim responded the winds were over 20 knots about 15 degrees to starboard. Ken and I decided to fly our approach to landing from left to right, directly into the wind. I would fly from the left seat. We completed the checklist and the crewmen in back strapped in, preparing for a possible bad outcome to this emergency.

Ken announced we were ready to land. We circled to the stern of *Milwaukee* at 150 feet and set up for about a two-mile final. The seas were calm and the deck steady.

Jim reported from the tower, "Your winds are now 15 to starboard at 24 knots. You have a green deck. Cleared to land."

We crossed the fantail for a no-hover landing in the circle indicating our landing spot. As I added power on our good engine, we almost were able to hover, but it took 100 percent power. I would have to land while keeping forward speed. If I tried to hover, we wouldn't have enough power and might drop onto the deck too hard.

I had to squeak off just a little power to settle in the center of the circle. We were down.

A few minutes later, Ken and I were in the wardroom having a cup of coffee and describing the failure to Jim. The ship supply officer wandered in and asked, "Do you guys know how many rigs you dropped in the ocean?"

We told him it was about 15 pendants and a bundle of nets. He angrily responded, "You guys owe me. That stuff all comes out of my budget, you know. Why didn't you just set it all back down on *Kennedy*?" I assumed the last was a rhetorical question. None of us responded to him.

It didn't take the mechanics long to discover the problem. While we were still in the wardroom, the maintenance crew chief came in and reported they found an actuator arm internal spring broken. They were digging for it in Supply as we spoke. They found the part and had it installed within an hour. By then, the other helicopter had landed, and *Milwaukee* would proceed to a destroyer in a picket station about 150 miles away.

Since the actuator arm is an adjustable linkage, we would have to fly a functional check to make sure the repair held. With a replenishment of the destroyer in the picket station scheduled at dusk, this would work nicely. We would fly the check flight about an hour before the rendezvous with the destroyer, then go right into the cargo transfer.

It was a beautiful autumn afternoon in the eastern Mediterranean. The sun hung low over the horizon, temperatures hovered in the mid-sixties, and about 12 knots of wind was blowing over five foot seas. We conducted ground hover checks and made small adjustments to the engine, then took off about 30 minutes before sunset.

Ken and I climbed to a safe altitude and methodically completed the engine checks in about 15 minutes. Everything worked well. The destroyer would meet up with *Milwaukee* in about 15 minutes, and they were scheduled to receive five or six replenishment loads from us.

In the meantime, the deck crew on *Milwaukee* moved the cargo into position on the flight deck and rigged it for five loads. As the sun descended below the horizon, the destroyer reported "green deck" and cleared us to deliver the first of the cargo loads. The ships were about five miles apart as we picked up the first lift. The winds were almost straight off *Milwaukee's* bow and the receiving ship was located about five miles to starboard, so I made the pick from the left seat. Ken made the drop to the destroyer's fantail just astern of the aft gun mount. The deck crew on the destroyer cleared the load from the small spot while we made the transit back and brought in the second lift, then did the same with each subsequent lift. With the ships slowly closing the gap, we completed the deliveries in 20 minutes.

It took the destroyer about five minutes to rig the retrograde for us to pick up and take back to our ship. Ken made the pick from the same spot on the destroyer just as the moon rose in the east. The ships had closed to within about two miles by now and steamed below us on parallel courses. I took the controls about halfway back to *Milwaukee* to make the drop. We planned to drop the load, then depart for about four laps around the pattern to update night landings. We set down the load and Ross, the crewman in the hole, reported, "Hook clear. Clear to go." I added power and dipped the nose about five degrees down to depart to starboard.

As we cleared the deck edge at about 70 feet above the water, the noise level abruptly changed, and we felt the rotor slowing. I was still keeping my eye on the ship's superstructure, working hard to clear the ship without a disaster.

"Shit, we've lost number two engine! Engine temperature is pegged."

Ross said, "Hole is closed. Secured aft."

Ken reported, "Ninety-three percent," indicating our rotor speed. "Fifty feet. I'm dumping fuel. Otherwise, we're going to swim."

"Go ahead with the dumps," I said.

We were clear of the ship now, and I focused on trying to trade off the little bit of altitude we had for enough airspeed to regain rotor speed and lift. The airspeed indicated zero, but we were headed right into the wind.

"Ninety-one percent. Thirty-five feet," Ken said just as the generators dropped off line.

We lost an engine, we couldn't climb or maintain altitude, and now we had no instruments or electrical power. Fortunately the evening was clear, we still had a nice crisp horizon, and I could make out the waves on the water. At least I could see to ditch the helicopter into the ocean. It seemed we'd soon be swimming.

Ken called, "Ninety percent. Getting worse, Kevin."

I guessed we were 15 or 20 feet above the water. I watched the waves, thinking I might have to pick the crest of one to settle on. But wait. There it was. The airspeed needle hopped. I squeaked off just a touch more power and tried to hold my height visually.

Ken said, "Turns are coming back a little. Milk it! Eighty-nine percent holding." Our rotor speed was increasing.

We stayed at that altitude for only an instant, but it seemed forever. I struggled to find just a hair more power and get the rotors turning a little faster without descending into the water. In the back of my mind I kept looking at the waves, trying to mentally match their rhythm if we had to set it down. Landing on the crest of a wave was our best option. Clipping a rotor blade into the water between wave crests could rip the aircraft apart and result in disaster for all of us.

"Ninety percent. Ninety-one percent. Ninety-two percent." Just then the generators came back on line with an audible change and a slight kick. I could see the airspeed now indicating 40 knots. I eased the nose up ever so slightly and started a climb.

Ken reported, "Ninety-six percent and holding. Passing 50 feet."

As we got to 70 knots, I took a little more power off and the rotor turns came up to about 104 percent.

Ken reported, "Hundred-fifty feet. Right engine is dead and hot."

We took a bit of a breath and climbed to 500 feet. I asked Ken to shut off the fuel dumps, which he did.

"Let's get that engine shut down." I rattled off the emergency memory items, and Ken concurred and performed the steps to shut off fuel to the number two engine. We alerted the *Milwaukee* flight control we had an engine failure again and asked if the ship could pick up speed and generate a little more wind for us to make a single engine recovery. In our condition, we had no hope of making a normal hover landing because more power was required to descend vertically than to fly forward. We had no choice but to land with forward speed. The ship had to move forward at that speed to give us a chance.

We went through the single-engine landing checklist while the ship accelerated. Jim was up in the tower again, and his voice came over the radio to tell us, "*HW13*, you have a green deck. We got your winds about ten degrees to starboard at 32 knots. How does that suit you?"

We said those would be great. We were lighter after dumping fuel, and with that kind of wind we might be able to hover over the deck. I started the approach, left to right again. I would fly a slowly decelerating approach to set it in the circle with no hover. Heck, we might not even drop the rotor rpms.

Emergency crews and firefighters were scrambling on deck. This approach had to be flown correctly. There was no room for error. Ross confirmed the two guys in back were strapped in and ready. Ken called out rotor speed and height above the deck. The other helicopter crew had launched and hovered to our right and slightly behind us, ready to pluck us out of the drink if we didn't reach the ship.

"Twenty feet," Ken called out. "Ten feet, Kevin. Looking good. You're right at the circle.

Five feet.

Contact.

We're on the deck, guys."

Half an hour later, Ken and I again stopped in the wardroom to deflate a little. The ship's commanding officer was alone in the wardroom, pouring himself a cup of coffee. The captain greeted us and said, "You two had quite a day! I guess you could call it a two-fer."

On the way into the wardroom, we passed the supply officer who asked, "How much of my budget did you donate to Davey Jones' locker this time?"

What a question. He was worried about the cost of nets and rigging, when we all know the real cost—and the only thing that matters in the long term—is the human cost.

* * * *

On that long ago October day when the American and French barracks were bombed in Beirut, I came face to face with the human cost.

After 11 hours of flying, my crew and I were running on reserve energy, powered only by adrenaline. We had no idea how many trips we made between task force ships and the destroyed barracks. We devoted every ounce of our time, energy, and prayers to flying rescue and recovery equipment to what we would later discover were bombsites. *Milwaukee* stayed on station with us for the day, also providing medical and damage control, and a chaplain to the beach.

We were directed to recover on *Iwo Jima* for one final trip as the sun neared the horizon over the Mediterranean to the west. We would carry an internal cargo load to the airport. The ship's cargo handlers and our air crewmen loaded and strapped down four pallets. We left the ship from a left downwind for our final trip, landing one more time beside the huge, smoldering pile of rubble.

Ross, being a normally curious air crewman, investigated the cargo while we were in route, to see what we were carrying. He stuck his head in the cockpit with an uncharacteristically somber expression. "Mr. K," he said, "these pallets are body bags—nearly a hundred of them."

CH-46 Sea Knight lifting cargo for Vertical Replenishment (VERTREP).

Directing the cargo drop at sea at dusk, Mediterranean

Deck crew during takeoff of CH-46

THE MILGRAM EXPERIMENT

by Ken Yamada

When my cell phone rang, I had no idea what was in store for me during the next twenty-four hours.

I'd been busy driving around the city from store to store, putting gear together before heading out to the Black Rock Desert for the infamous alternative generation party called Burning Man. Five days of drugs and sex awaited me, along with a sea of twenty thousand young, horny revelers. This was the party's second year in the Sierra Mountains on a dried up lakebed, in the borderlands between Nevada and California. It was a dry, dusty and horrible place to live, but ideal for five days of lawless debauchery.

I'd been looking forward to this party for months. Stuck between a job I didn't like and parents who were growing old, I needed an unthinking man's vacation.

So there I was the day before the festival, driving around San Francisco picking up what I needed for the five-day camping trip. I already had a tent, sleeping bag, and a butane cooking burner. I needed plates, forks, knife, water, a flashlight, batteries, suntan lotion, a hat and food. I tried to consider items that would survive in the desert for five days. I bought bread, rice, pasta, olive oil, salt, cayenne powder, salami, sausages, cheese, a can of peaches for a desert dessert one night as a treat under the stars, a can of green beans, three cans of tuna, three gallons of water, and bananas.

With gear and supplies in tow I headed back to my apartment for the last night of good sleep before the festivities. That's when my cell phone rang—beginning the real adventure. The number was unfamiliar but I picked it up anyway, thinking one of the guys had gone up early and needed something. A vaguely familiar voice said "hello," and slowly it dawned on me he had an Italian accent.

And then it hit me. He was a friend of a friend, named Antonio.

"I just got in from Rome," he said

I searched bits of memory from a night of revelry four years earlier. We were running around Rome with an Italian soap opera star named Mimmo Lori and ten young giggling Italian girls in tow. Like hyenas with sick prey, Antonio and I picked off any girls who strayed too far from the celebrity.

"I just got in from Italy, and I don't know anyone in this city," He said. "What are you doing? You want to hang out?"

My next words tumbled out of me on their own. "Hey, do you want to go to a massive party in the desert?"

As I headed out to the airport, I wondered if I'd even recognize Antonio. I knew he had dark brown hair, was about six feet tall, and moved with a nervous jitter.

When I pulled up to Alitalia, an Italian version of Tom Cruise flipped a duffle bag over his shoulder and headed for my car. He wore a biker's snug leather jacket, tight euro style jeans, and leather sneakers that could have doubled as working shoes. He still had the nervous energy around him. A swatch of windblown hair kept falling in front of his eyes—eyes holding the same beady-eyed intensity Tom Cruise gets when Oprah asks him if he's getting divorced again: A look that makes you wonder how many marbles he's missing.

Antonio got into the car, and we hugged and kissed each other on each cheek.

"Geez, it's been a long time," he said looking me up and down. "You been working out?"

I wanted to lie, but considering how fat I was, he wouldn't have believed it anyway. "No, not really," I said, much quieter than I intended.

"Yeah, I can see. I didn't want to say anything but you look fatter, like a butcher."

We laughed about this for like thirty seconds, twenty-nine seconds of which I faked. The entire time thinking, *I can't believe I'm going to be stuck with him for another six days.* I was suddenly grateful for the bag of "Purple Urkle" marijuana I'd bought the day before.

"So how are we going to get up to this party?" He asked, getting right to the point.

"In this car," I half-heartedly responded, still wondering if he had an aunt or an uncle he could stay with.

"Driving? That might take long, no? Let's fly?" He said this so matter-of-factly it almost sounded rational.

I'd read about a few daring people who flew into the party and landed on the flat lakebed. The San Francisco Chronicle featured photos of Indiana Jones types standing beside a plane, surrounded by ogling women.

Unfortunately, I had John Madden's disease, contracted a few years before while flying home for Christmas in San Francisco. I was with my good friend Matt, a longtime companion and one of the original founders of the Pyramid Camp up at Burning Man.

We were flying over Denver and the "fasten your seat belt lights came on." As always, I ignored it. Besides I was engrossed in a competitive game of Gin-Rummy. Suddenly, we hit an air pocket. Although I'd spent much of my childhood on a plane following my Dad around the world, I'd never experienced anything like what was about to happen.

Our plane free-fell for the next ten seconds. I floated out of my seat until I was three to four feet above it and had to push off the baggage bin above me to stay within shouting distance of Matt. Our game of Gin-Rummy rose into the air around us; some sets of cards staying together, while others hung in the air at different heights. Matt and I and about two hundred other people started to scream.

Ten seconds passed, and we managed to shout young male—just before you die—type deep comments to one another like, "I knew I was going to die with you, man!" "Fuck!" and "Holy Shit!"

The next thing we knew, the plane bucked, shuddered, and seemed to regain its gravity. I came slamming back down onto the arm of the chair. My bowels felt like I had to release a river of diarrhea, the cards fluttered down around us, people began whimpering and praying, and a lone woman screamed at the top of her lungs. The ceiling lights flickered on and off, a few oxygen masks had fallen down and were swinging wildly, and then babies began to cry.

A few minutes passed and the pilot came back into the passenger cabin to give us a speech about how we were never in real danger—even if it felt that way—and we'd soon be landing, so no worries.

For the remainder of the flight, we all wavered between exhaustion and wide-eyed fear, resembling a group of hostages returning from captivity.

Strangely, my fear of flying didn't develop right away. Like a disease, it took a full year and a few more flights before the fear seeped into my consciousness. And when it began, it managed to wrap around my throat, until I would get cold sweats the night before a flight, and they didn't stop until the seat belt lights turned off.

In this state of mind, I confronted the thought of flying up to Black Rock Desert.

"Hold on. Now wait a second. How are you planning on getting a plane?" I asked. Just to name one of the plethora of problems I foresaw. "Do you even have a pilot's license?" I asked incredulously.

"Of course," he said. "Why? You're not afraid, are you?"

Instantly, I felt twenty pounds heavier and images of Gene Wilder running around in a chicken suit with Richard Pryor sprang into my head.

"I've flown solo ten times." He said, showing me all his fingers on both hands as if to eradicate any doubt about his ability to land a plane onto an illegal and non-conventional runway somewhere in the Sierras. This filled me with greater fear, because it all sounded too simple—like he'd seen too many episodes of Mission: Impossible. I wasn't so sure he understood we wouldn't get a second take.

"No, Dude. I don't think so," I said, unable to think of any good arguments.

Suddenly, I remembered an article in the same Chronicle just after the one about the hero types who flew in on a plane. An article about five friends, sound asleep in their sleeping bags, run over like five squishy burritos by drunk revelers racing around in a pickup truck at night.

"C'mon, it will be so much faster by plane. It's cheap to rent and they only charge you by the miles you fly, not the time. I've had my pilot's license for almost a year now. Which in Italian could mean "more than a month."

"No, we don't even know where to rent a plane. No!" I tried to sound final.

My phone rang, to my relief, giving me what I thought was a reprieve. It was my good friend Matt calling from Burning Man, sounding panicked as he blurted into the phone, "Dude, they have checkpoints. They're looking for drugs."

I wasn't too worried about this, since I only had marijuana, which I could easily hide in my backpack.

"No dude, it's a fucking major problem. We got like two hundred hits of E we were going to bring up. If they find that, we're going to jail for life," he continued.

"Well, I'm not sure why you're calling me," I said, knowing what he'd say next.

"You got to bring the drugs," he said. "Borgo doesn't want to bring it. He's bailing on coming up."

I looked over at Antonio, pacing back and forth, never able to hold still, just far enough away it looked like he was giving me some privacy, but I could tell he was trying to listen.

I mumbled, "No way. I don't want to go to jail for life either. Forget it."

Antonio sauntered closer, sniffing out correctly that something was brewing. Even though I was mumbling, trying to play it low key, he could smell blood in the water.

"Come on! People are counting you. That's gonna be our main fun for the next couple of days," Matt half whimpered. "Otherwise, what are we going to do?"

In the background, as if on cue, I heard some girls goading me on to do it for them. "Come on, Ken, do it! Come on, do it!" I was wrong about Matt's English being too difficult for the crazy wannabe fly-boy Italian to understand. I guess "Bring Drugs," is easy in any language.

I got off the phone even more depressed then before. I looked at Antonio, and contrary to the actions of most people who are about to traffic drugs and fly unsafely, he seemed excited. I was fighting a losing battle.

"Cheer up my friend. You see, it's our destiny," he said, slapping me on the back.

The following morning I awakened with a dry throat and a knot in my stomach. I vaguely remembered playing a starring role in three separate film nightmares—*Alive, Midnight Express* and *Twilight Zone.*

I looked over at Antonio, still asleep. He lay there snoring, one leg precariously perched on the edge of the bed. I'd never really noticed how much hair he had. It bordered on fur. I wondered if this was some sign of primitivity, hence stupidity. Australopithecus, Cro-Magnon, and Neanderthals were hairier, after all. Can an ape fly a plane I wondered? And then a faded 70s Technicolor picture floated into my head: the poor face of the space monkey right next to all those buttons in the cockpit, looking confused, irritated. With this image in mind, Antonio no longer had a swashbuckling air about him. Instead, he looked like a T-shirt-wearing ape that got knocked out and thrown onto my bed. For a second, I thought about bludgeoning him to death. Then he opened his eyes. Damn.

The office for the flight school at a small airport on the outskirts of Oakland looked like a rusty storage shed. It was a modest, one-story wooden frame structure, with corrugated tin walls and roofs. A dilapidated screen door never quite closed all the way. It surprised me when Antonio said we could rent a plane. Somehow, this sounded like the equivalent of renting a gun for the day, bullets and all. It seemed a bit cavalier for something so seemingly dangerous and complicated. You can rent planes by the hour, mileage, or by the day, almost like renting a car.

"So, do you have your pilot's tools and California and Nevada area maps?" The lady at the unofficial front desk asked. Antonio looked at her blankly, then laid on his thickest Italian accent and said he left them in Italy, because he hadn't expected to fly in the United States.

"Oh, you're from Italy?" She said. "My husband and I went to Rome and saw the Sistine Chapel. We loved it."

Then Antonio did his movie star smile and said, "I'm from Rome. Next time you come, you must call me. There are some things in Rome, only a Roman knows."

My body almost convulsed, and I nearly fainted with fear as his Euro-Trash-Jedi mind trick made her fingers sign the release papers. I wondered how many other people she'd signed off to their deaths. And before I could say "pasta," Antonio was herding me out of the office.

This part of the airfield looked like a forgotten hanger on some outer section of a large international airport you see as you taxi down the runway before take off on a real plane. Empty warehouse-type buildings sprouted unused concrete runways with cracks and foot high weeds growing out from them. Layers of faded painted lines that once guided pilots now acted like decoys, taking you off the runway. It was easier to ignore them and guestimate rather than follow the lines.

A fit, medium-sized white man with a mustache, wearing butt hugger khaki shorts and pilot's Rayban sunglasses greeted us outside the office. He looked like a retired member of the Village People, or an Italian.

He looked Antonio up and down hungrily and said, "Well, I guess it's going to be just you and me, partner. I'll be the one testing you to see if you're ready to rent the plane."

Antonio smiled, pulled his jeans up higher, and responded, "Well, I hope I can live up to your expectations."

The instructor twitched happily, and off they went together like two people plotting how to kill me. Antonio turned around and winked at me.

I turned away and looked around. A group of four Cessna single-engine props huddled together on a patch of the tarmac. This was the first time I'd seen a single engine prop plane so close. I walked over for a closer look, imagining how quickly my death would occur. One was an ordinary white Cessna 172, a single engine, four-passenger, five hundred miles of flight on a full tank, propeller plane—like a go-cart version of a real plane, and not much longer than a station wagon. The windshield and headlight were about the size of those on a car. You could open the four windows around the cockpit manually. The thought of being able to open one of them while ten-thousand feet in the air flying at one hundred and fifty miles an hour made the cockpit seem vulnerable. The seats were simple hard plastic bucket seats, like you'd find on a cheap old roller coaster ride, and the plane must have been built in the 80s, because the paint design reminded me of a 1983 Fiero with wings— cheap and cheesy. If this plane had a head-on collision with a Honda Civic, the Civic would win.

I looked up and studied the plane Antonio was flying. It landed a few times and took off a few times. For a brief moment I was impressed to see a friend actually flying a plane.

I noticed a large black man in mechanics overalls working on a nearby plane. He periodically looked up to watch Antonio's plane. I walked over to him.

"Is that the actual plane we're renting?" I asked, pointing to Antonio's plane.

"Oh no, no, no. That's the flight instructor's plane. You'll be taking that one over there," he said with a heavy southern drawl, pointing to a plane sitting all by itself, looking a decade older then the planes I'd been looking at. "I just got her ready for you," he said with loving care showing in his expression.

"Oh. You mind if I take a look?" I asked.

"No, go right ahead," he said.

I walked over and immediately noticed the windshield had hairline fractures in it. I ran my fingers over the glass, thinking they must be deep and the surface would still be smooth. I was wrong and horrified as the little cracks and uneven platelets of glass scraped my fingers. Disturbed, I looked into the cockpit and noticed the hard plastic of the instrument dials were cracked, making some numbers even difficult to see. The bucket seat enamel was also cracking, the seat belts were slightly frayed, and there was some random junk in the back of the plane—buoys, fishing net, life jackets. I walked around to the front of the plane. A chunk of metal the exact size of half a golf ball was missing from one of the three propeller blades, leaving an open gash. Maybe it was the images of the shards of glass from the windshield plowing into me at two hundred miles an hour, or the plane spiraling wildly out of control because of a non-aerodynamic propeller, but I suddenly wanted a different plane.

"Hey, do you think this plane is okay to fly?" I shouted over to the mechanic, suddenly more worried than before.

"Oh, you mean Charlotte," he called the plane. "Yeah, she'll take you anywhere. She's been through it all. She's got plenty of experience. Besides, I just put a new engine in her last month. She's one of the best in the yard. Just tuned her yesterday. You got nothing to worry about."

"What about the windshield? It's got like a ton of hairline fractures in there." I hoped that would tip the scales and convince him to give us another, better plane.

"Where you guys flying to?" he asked.

"Up to Reno," I said, careful not to mention Black Rock Desert.

"Oh?" he said, seeming concerned. "Your buddy been flying for a long time? He know this area well? It's not easy getting over the Sierras."

"No, no and no. I don't think so. Why?"

"Well, you know he's gotta get some altitude early. Can't wait till he's at the mountains. Won't have time to climb to the necessary height in all that thin air," he said, his left arm mimicking the Sierra Mountains, his elbow the peak. His right hand playing the part of our little plane trying to fly over his left elbow and just doesn't make it, crashing into the side of the mountain. He notices me looking at his visual demonstration in horror.

"Why can't we take one of those?" I pointed to the newer planes in the group.

"Oh, those are for students. You'll be just fine. Charlotte won't let you down." He turned away from me, not wanting to cause trouble now that Antonio and the instructor were coming back.

Antonio climbed out of the test plane with his instructor in tow. "He's pretty good," said the smitten instructor with a glazed look like Antonio had just given him a blow job in the air. "I think you guys should be fine. I'm going to clear him to rent the plane."

After we signed the documents, including waivers that intimated, "If you should die while in one of our planes, no matter who is flying it, no matter if the windshield shatters embedding millions of pieces of glass into your body, or if the golf ball size hole in the blade sends you flying in the wrong direction, it's your fault because you were stupid enough to rent this plane."

The instructor also handed Antonio an extra map of the California and Nevada areas and followed it with a wink. "Here, use mine," he said and pressed the map into Antonio's palm.

We packed our goods into the plane. Every item felt like a heavy shovel full of dirt being dug for my own grave. I wondered if I should have packed the "E" in a different section of the plane but figured we'd be dead anyway. By the time the cops found our burning wreckage, they would already know it was yet another failed drug run.

Now, minutes from embarking upon what I now feared would be my last flight, sitting there just off the runway, Antonio perused the three-page checklist of equipment and instrument checks faster than I could read them. "Landing lights, safety light, electrical gauge, gas cap…" He was definitely not doing the checking. "I think we're okay," he says after about 30 seconds.

"Um, aren't you supposed to get out of the plane and look at some of the things written down on the check list?" I asked. "Maybe flip some switches, tap some gauges? What the fuck man? I mean we aren't driving a car, we're talking about flying."

Begrudgingly, and more for my sake than his, he started flipping switches, at times seemingly knowledgeable about them and at other times surprised with the sounds and lights some of them produced. He periodically muttered technical terms—electrical, tail rudder—every now and then, as if to prove he knew what he was doing. He even got out and looked at the propeller and opened some hatches, all the while looking at me from under his mop to see if I was satisfied yet so he could get back into the plane.

When finally he taxied the plane down the runway, it felt light, like we were driving in a cheap car, going at top speed. It didn't feel like it was going to get enough speed to even take off. I'd ridden in cars with more heft and pick up than this plane. And, it was unbelievably noisy. I kept wondering how something this small and flimsy and powerless could get over the Sierras. One yard, ten yards, twenty, thirty, a hundred and suddenly the ground fell away from us and we were in the air.

I measured our height by how quickly we could die. Ten feet, safe. Twenty feet, near death. Then, a hundred feet, death for sure. It felt like some Ferris wheel ride at the carnival, sitting in a bucket being hoisted hundreds of feet into the air, the pull of gravity, and the rush of going higher and higher until the ground below looked surreal, like some large bird's eye view photo of the Bay Area. To be honest it wasn't as scary as I imagined.

Only when I looked out the windshield to where we were headed could I see the faint outline of the Sierra's, like distant storm clouds on the horizon, awaiting our arrival. But for now, I actually enjoyed the feeling of flying.

"BBBBBBBBBBBBB…," the engine was the only sound I could hear. We had to shout to one another to be heard. That's when Antonio pulled out an extra pair of headphones from his vinyl flight bag and handed them to me. "You see, you're lucky I remembered to bring this for you," he said proudly.

They looked like old 70s DJ headphones with a thick ring of foam around the ears and a wire mike to speak into. I wondered how he managed to remember to bring an extra pair of headphones, but not the necessary navigation equipment.

I looked inside his bag to see what other "important items" he managed to remember to bring other than his navigational tools. A few packs of chewing gum lay on the bottom, pens, a flight logbook, a rag, a baseball cap, and a ton of receipts. It looked like a bag belonging to a 13-year-old kid. Proudly, he slapped me on the back and rocked back and forth in his seat, unable to contain his excitement and said, "Pretty cool huh? You can hear the tower and we can talk to each other without yelling."

It slowly dawned on me that conceptually Antonio still thought of flying as I would a trip to the beach. Fun, whatever you bring is a bonus, and whatever you forget you can do without.

I plugged the headphones in and instantly I could hear the Oakland airport tower talking to the planes in the air. Slowly, I started to understand what they were saying and what it all meant. The tower would call out a series of names and numbers, which signified which plane they were talking to. "UA875 you are fifth in line to land, looks like it'll be at least another fifteen minutes."

Calls like this went on for quite awhile, with different airlines and flights. It all fascinated me to be able to listen in to the "official," sounds of the pilots. I was getting pretty good at understanding these calls.

"That's pretty cool listening in on the tower," I said, starting to enjoy myself.

As if my words jarred his memory, and as if I wasn't allowed to relax, he went from staring dreamily out across the landscape to suddenly turning something on called the "transponder." I later learned this was the instrument that allowed the air traffic controllers to see and identify us on radar.

Antonio opened the map the instructor lent him and casually steered the plane with his other hand. Watching him pretend to deeply understand the nuances of the map was akin to watching a dog play with a calculator. I could tell he neither understood the letters nor the shape of the topography. He became increasingly frustrated as he looked at the map, then at the terrain below. He pointed to a lake on the map shaped like a Frisbee and said, "Does this look like the lake down there?"

"No," I said shaking my head in disbelief, wondering how he had made it from Italy to America. "I think that lake down below is this lake." I gently pointed to the correct lake, not wanting to frustrate or frighten the caveman in charge of our lives at eight thousand feet above the ground. The lake I pointed to was on the opposite side of the airport.

He quickly flipped the map around 180 degrees and inspected it again. And he also turned his frown upside down and into a creepy little smile.

"That's better." He said. "Hey, you're pretty good at this. Why don't you be the navigator and I'll fly the plane."

Once we figured out our general route, he asked if I wanted to fly the plane. Ordinarily I would have said, "no" but seeing how he'd performed for the last few hours, I said, "yes," determined to learn everything I could, figuring the less involved he was in this process the better chance we had to survive this adventure.

So he showed me how the throttle worked. "Push it forward when you need power, for like when we're taking off or to climb."

He put his index finger on a round dial. "The altimeter shows you how high we are above sea level." Tapping an instrument in front of him, he said, "The artificial horizon shows you how level the plane is. And this transmitter allows the tower … oh yeah you already know," he said, and we both laughed.

I grabbed the wheel on my side of the plane. He flew with me in the beginning, still controlling the plane with his control yoke, like a father would with his child and a fake child steering wheel. Surprisingly, it wasn't so hard. Each movement of the plane had some inertia to it, meaning if you steered right it would go right and then some, which made you have to under steer and be patient with the turns, but as long as you kept it steady, the plane pretty much flew itself.

I took her up and down a bit and banked her left and right. I imagined how WWI pilots must have felt while flying their propeller fighter planes. I was excited to do something so different from my ordinary life. Antonio grinned at me like a proud father, and squeezed my shoulder, and said, "Look, you're flying this plane," and I finally loosened up.

Antonio opened the window and stuck his hand out, surfing the passing wind with his hand. We laughed like kids playing hooky, and for just a little while I forgot about life and death and began to have fun.

"Hey, man thanks a lot," I said suddenly feeling bad for all my earlier doubts. "I would never have learned all this without you." I was excited to stretch my wings and was grateful Antonio had pushed me out of the nest. And lake by lake, landmark by landmark, we flew connecting the dots on the map, using highway 80 as our yellow brick road to slowly snake our way towards the distant Sierras.

Our fun was interrupted by the plane's radio, "November 9-4-1-5?"

I looked at the brass plate on the dashboard. It read, "N9415."

"November 9-4-1-5?" called the tower again, this time in a more steely tone. "If you are still on Oakland Tower frequency, turn left immediately. Contact departure control on 135.55 now."

There was a sense of urgency in the radio call. I looked over at Antonio and wondered why he wasn't responding. He was busy scribbling something into his flight log book that rested on his lap. A loaf of bread about the size of a football protruded from his mouth, and in his free hand he held a log of salami. He had no idea the tower was calling our plane.

"Hey, dude," I said. He kept scribbling and chewing and gnawing. "Dude! I think they're calling our plane." I shouted out to him in case he couldn't hear over the intercom headsets.

He looked over at me, and then at the dashboard where the plane identification was. And like a parent cleaning up a dirty child's diaper he wrinkled his nose. Nervously and slowly he picked up the mike. "This is flight N-9-4-1-5," he said slowly, his finger pointing to each number and letter as he said it.

"Is there a reason why you weren't responding?" The controller asked.

"No I was busy with something else," Antonio responded a bit rebelliously like you would to your angry older brother.

"Where are you heading flight N-9-4-1-5?" The controller asked again, slightly annoyed by Antonio's tone.

"We're headed up to Reno, Nevada," Antonio said, almost giggling, and gave me a knowing wink and pointed to the mike and mouthed the word, "asshole."

The controller responded coldly, "Sir, please respond with your call sign first. And Reno is North East from here. Your heading is N-o-r-t-h E-a-s-t. You are flying directly into a temporary flight restriction area. Turn left now. Disregard the frequency change to departure." He spoke slowly and condescendingly, the way a traveler might try to communicate with someone who only understands a different language.

Antonio then repeated back to him in a mocking impression of a mature voice. "Yes, sir, that's right. We're headed northeast, but I am turning left now," turning the map left, then right, confused and trying to comprehend what the controller was saying. I could tell Antonio, like me, had no idea what a temporary flight restriction area was, nor did we know such a thing was anywhere nearby. I guessed flying into one was a bad idea.

"November 9-4-1-5, please monitor guard frequency on 121.5. Good luck, gentlemen."

"Yes, we copy. Thank you. Over." Antonio half mocked him, promptly put the mike back on its hook, and sighed with relief, like he'd just finished with a prank phone call.

"You're supposed to repeat your call sign back to me," said the controller, sounding more annoyed. I expected the controller's next question to be "Where did you buy your pilot's license?" He was a professional, though, and I suspected humor was not warranted under these conditions.

Antonio quickly picked up the mike again and keyed the switch with his thumb, "Oh, sorry we don't do that in Italy."

"Good luck gentlemen," said the controller again, glad to be rid of us.

"Pfff, good luck." Antonio mumbled annoyed.

It was like witnessing a spat between two fifth graders, except this one was flying our plane.

Of course, Antonio didn't seem to notice. He was already fidgeting with the radio, trying to get it on to the new frequency.

"Good luck? That's not good is it?" I said, trying to draw his attention to some important details about flying etiquette. I figured we needed all the support we could get from these guys in air traffic control.

"Shhh, I can't hear," Antonio went back to fidgeting with the radio.

We flew on in relative silence for about ten more minutes, the static from the new radio frequency drowning out conversation.

Antonio looked ahead, down at the ground anxiously, and I followed his gaze out the window. In front of us loomed Sacramento, like a gray stain on the earth.

"You know what's weird?" he asked.

"No" I shrugged, exhausted and wondering what a nightmare it must be to date him. "What now?"

Slowly, confused, he muttered, "We're really close to Sacramento, but I can't hear the tower. I wanted to listen to the tower in case there was other traffic around here." He tapped his headphones and turning the radio dial back and forth rapidly out of frustration. He pointed to his map and the Sacramento airport frequency printed inside a little box.

I look at the radio, and it seemed odd to me, like coming home and feeling like somebody looked through your stuff, but nothing obvious seems really out of place.

"Hey, I think the last two digits of the radio are reversed," I said. "Switch the last two digits on the radio."

He looked at me, annoyed, and stubbornly ignored my comment and didn't try it for awhile, clicking on the mike a few more times and even trying Oakland's frequency one more time. The Oakland tower sounds faded in and out between crackles and static as if saying goodbye. Finally, Antonio tried the frequency I suggested.

The moment he clicked the radio digits into the correct order, a voice came in loud and clear: "Alaska 3-4-5, be advised we have reports of turbulence on your route, change to departure now."

Antonio looked at me sheepishly and shrugged. "I switch numbers around sometimes."

Oh, great, I thought, *he's also dyslexic!*

"Hey, how many hours have you flown? For real?" I asked.

Antonio shifted in his seat nervously and muttered, "It depends. In Italy or in the US?"

I looked down at the logbook between us in the bag. "So, who checks your logbook to see if it's accurate or up to date. Or legal?" I asked.

Antonio suddenly looked at me suspiciously and tried to zip up his bag, but the zipper kept getting caught. "In America the FAA checks every now and then, but in Italy they don't check that much. It's more on the honor code there."

"The honor code?" Somehow this horrified me. Italy and honor code didn't seem to fit comfortably in the same sentence. "What! You mean like you could write whatever you want? Well, how do they know you're not making it up?"

He seemed indignant now. "They check in Italy, okay?"

"Who checks? Is there like an official body like the FAA in Italy?"

"No not really, the guy who checks is a family friend, okay?"

No, it was not okay. I'd lived in Italy a few years back, and when I needed to extend my student visa to stay longer than 90 days, I was told I needed to get a signature from the postal office first, and then one from the train station after. So I first went to the post office. When I got up to the window, the lady shook her head and said I needed to first go to the train station and get a stamp from them before she could sign off on it. So, I then went to the train station, but the man at the window shook his head and said I needed to go to the post office first and get their approval for him to give me the stamp. That was Italy.

I reached into his backpack and pulled out his logbook. He tried to grab it from me, almost tearing it in half, but I pulled it away from him. The plane veered off to one side, and he was forced to grab the control yoke instead.

I opened the logbook as he stared menacingly at me and periodically tried to get it out of my hands. Each page was divided into 6 columns:

Column 1 had the date.
Column 2 indicated where the flight started from.
Column 3 where the flight flew to.
Column 4 indicated what time the flight left.
Column 5 indicated what time the flight arrived.
Column 6 indicated how long the flight was.

I remember cheating once on a homework assignment when I was in 5th grade. I was supposed to keep a journal for a month, and I didn't. So on the day it was due, I sat in front of my locker and started to make up entries for every day of the month. Next to me was another smarter boy in the same class doing the same thing. Except, he had four pens in his hand and switched off between them every few entries or so. At the time, I thought nothing of it. In fact, I thought he was silly and paranoid. Needless to say, I was busted for cheating and he was not.

I looked at Antonio's flight logbook. The first 20 pages were comprised roughly of six months and had a lot of sporadic entries over various dates. The pen colors and thickness were different; some even looked as though they were written in by someone else, maybe an instructor or official with whom he flew. But then his entries changed. The next 20 pages were all written in pencil—the same pencil—with a haunting similarity to my fifth grade journal. I saw many more entries in pencil than in pen. Each penciled-in entry was for longer flights than the ones in pen and ... to make matters worse, I noticed six or seven entries on every pencil page were erased and written over again to include more hours.

I went back to the beginning and counted the number of hours he had so far. He only had about 100, including the *bonus pencil hours*. At least seventy were *pencil hours*.

"How come so many of these logs are penciled in?" I asked.

"Oh, I just wrote those in with pencil so I can write them in later more neatly with a pen." He said.

I could tell he was lying.

"How long do you have to fly before people consider you an experienced pilot?" I asked.

"Maybe like 400 hours."

"How many hours do you think you've flown?"

"Like, 400," he said.

"Well, actually you're wrong," I challenged. "You're not even a quarter of the way there, even including your penciled-in hours. If you minus out the penciled in hours it's like you're not even a beginner. I just counted it up, and you have 100 hours and most of them probably are illegal. Dude, that's really different from 400."

He looked at me sheepishly. "Relax. I needed 150 to apply for this job, otherwise they wouldn't even consider me, so maybe I added a few."

"Added a few? What the hell kind of answer is that? Do you realize you have MY life in YOUR hands? I mean, dude, how many hours have you flown for real?"

Antonio rocked back and forth in his seat uncomfortably and bit his lower lip. "Like by myself?" He asked looking at me sideways.

"Yeah," I nodded, somewhat frantic now.

"Maybe 40 hours," he whispered, "but you can't tell anybody, okay?"

I felt like smacking him. "Forty fucking hours? That sounds like how much time it takes to learn how to ride a bicycle, not a plane. Does forty hours qualify you to land on some random lakebed with this thing? Have you ever flown over mountains before? You're already sort of lost and screwing up your radio conversations. I mean, have you ever even flown in the US before? We got to turn this thing around. We've gotta land."

I reached for the control yoke and started to turn the plane around. He fought back and kept the plane flying straight. We battled over the direction of the plane for about a minute, with the plane swerving downward dangerously.

"What are you doing? Turn this thing around!" I said. Knowing in the pit of my stomach there was nothing I could do to force him to turn back. I was trapped.

He looked at me defiantly. "No." He grasped the steering wheel. Now I knew how Eva Braun must have felt in the bunker, stuck with Hitler during the last few days of the war.

In that moment, I felt the full brunt of his deceit. It had nothing to do with deceiving me. Antonio was so stupid he was deceiving himself. I just happened to be a hostage to his idiocy, helplessly along for the ride.

Trying to change the subject and shake off my menacing stare he shouted, "It's too late anyway. We're already at the mountains. Look," he pointed out the window in a panic. "We have to gain altitude, I think we're too low. Help me find the lowest crossing over these mountains," he shouted at me, throwing me the map and putting us in scramble mode.

I didn't realize we'd been arguing so long, but there before us loomed the snow-capped Sierra Mountains—so close I could see details in the ridges and individual trees.

I looked at the altimeter; we were at 8,000 feet and climbing. I looked at the map and saw most of the higher peaks were around 9,000 feet.

Antonio was busy pulling on the steering wheel, trying to make sure we had enough clearance, pulling the plane straight up—we could only see the sky. I felt the G-Forces pulling us back down. My adrenaline started pumping, and I completely forgot about turning the plane around. Instead, I focused on the map and seeking a way through the mountains.

The map was useless in the mountains unless you were experienced and knew the terrain or had mapped a course beforehand. We had no more highways, lakes, and landmarks to guide us. The only way to tell the mountain peaks apart was by height, and although on the map each mountain had numbers indicating how high they were, I couldn't tell which peak on the map correlated to the spot below us. We picked the highest mountain in the area, Squaw Peak at 9,000 feet and decided to fly toward it and at least two thousand feet above it, so at least we'd know our location.

Soon we were well into the Sierras, flying over the mountains, and although some peaks still seemed high enough to clip us, having gotten used to seeing the ground eight thousand feet below us, I felt relatively safe flying high over the ones around us. We went in and out of puffy little clouds, and Antonio even slalomed between a few, like in a video game. For awhile, I was mesmerized by the raw act of this tiny, flimsy, non-pressurized, single engine plane flying high, taking us over the Sierras. I studied Antonio as he gazed over the mountains in awe. In some weird way, this was empowering. I forgot about turning around. I got a second wind of confidence and suddenly felt it was our destiny to reach Burning Man, landing like Indiana Jones on the desert floor and reaping the fruits of our coolness. I thought of the bikini-clad women who'd greet us.

When Antonio turned to me and spoke, I realized something was wrong. His speech was slurred, and he wore a goofy smile on his face. He kept rubbing his eyes and smacking his dry lips and saying "I'm really thirsty. Can I have a bottle of water?" I reached for the water in the seats behind us and that's when I noticed my head was also throbbing, my vision seemed blurry, and the inside of the plane felt very cold. I looked over at the altimeter and noticed we were at 19,000 feet. Somehow between our panic, awe and upward gusts of wind we'd accidentally flown up to 19,000 feet.

Antonio was now blinking slowly, his eyes sometimes remaining closed for a few seconds at a time. Between short breaths of rarified air, I managed to yell at him and point at the altimeter. He looked at it, made the thumbs up sign to me, and continued onward at the same altitude.

Again, I pointed to the altimeter, shook him this time and yelled, "We have to fly lower."

He slurred back to me, "Um afraith tu fly lower, I don know how high the peaks are."

We argued in slow motion for the next ten minutes as my head started to thump and his speech got even more slurred. "Dude, we have to go down. We're gonna pass out at this altitude," I yelled. "We need oxygen, for God's sake. Take us down." I later learned half of the oxygen in the atmosphere is below 18,000 feet. We were at serious risk of dying.

Finally, all the while saying, "O-keaaay buth I I I don thinnnk iths a goo idhea," he gradually lowered the plane to 14,000 feet. And like poison passing through our system, our minds cleared and we were able to slowly regain our wits.

We passed way over Squaw Peak and headed in the general direction of Lake Tahoe, an area familiar to me and on the way to Black Rock Desert, hoping somehow to catch our bearings from there. Other than a crude reckoning based upon the setting sun, we didn't even know what direction we were heading. We flew like that for ten more minutes, and like a long lost lover, Interstate-80 peeked out and winked at us between the mountain passes and ridges. Again we followed our yellow brick road, like a siren song, beckoning us forward, and deeper we went into the Sierras.

Finally, landmarks appeared—mainly the Truckee Tahoe airport.

"Hey, I need to go to the bathroom. We got to land here," Antonio said pointing to the small airport below us.

"Just piss in a bottle," I said.

"No, I have to take a shit." He said, looking at me pleadingly.

We called to traffic on the radio at Truckee Tahoe, but no one answered. So we landed at Truckee Tahoe airport and taxied the plane just off the main runway. Our position was far away from any functioning building with a bathroom, so Antonio ran behind a shed near our runway and did his business on the tarmac. I was his lookout.

Strangely, no one ever came or said anything. The airport seemed abandoned.

We took off again from the empty, silent airport, and headed in the general direction of Black Rock Desert. As we cruised at 12,000 feet, I could see the sun setting northwest of us about to fall behind some mountains. And then it dawned on me we were running out of sunlight. "Hey, how long do you think we have before we get to the lake?" I asked him.

"Maybe like 20 or 30 minutes," he said.

I quickly looked at the map. Black Rock Desert was north east of Truckee, and although it didn't look all that big, I figured with 20,000 people attending, we were sure to see the encampment. I studied the map again, but feeling tired, I closed my eyes and fell asleep as we flew in silence toward Black Rock Desert.

I awoke to the familiar sound of the single propeller engine droning. The sun had disappeared behind the mountains, and we were now flying in the ambient light reflected in the sky. At most, I figured we only had another 20 minutes of light remaining.

I looked over at Antonio, who was busy looking out the windows at the ground below. By his desperation, I had a sneaking suspicion we were lost again.

"Hey, aren't we supposed to be there by now?" I asked.

"Yeah, yeah, we'll be fine. We've only been flying for like 30 minutes. We should be there soon. I'm just trying to find the lake," he said. He handed me the map. "You can help me by looking for it on here," he said. He continued his frantic search.

After about ten minutes it became too dark to see anything on the map. I looked across the landscape, where the beautiful mountain range with cartoon-like snowy ridges with ski runs and fancy inlets of homes nestled in the mountains, were being washed away by the long shadows of the coming darkness, disfiguring the landscape and taking away any we could use. Our confusion about where Black Rock Desert was located in this dissolving landscape was replaced by blindness.

An unspoken realization crept over us, and like two practiced athletes who know when the match is on the line, our brains went into overdrive. We looked back and forth between the map and the terrain, trying to find some matching, recognizable feature. Instead, we met a growing sense of dread. Within minutes, the ambient light went away and we were flying in complete darkness, the land below us a black abyss.

Somehow, I'd never imagined flying in the dark, much less over the Sierras. Somehow, until that moment, I'd never considered how much thinking and foresight our society had already done for us. In the air, in uncharted territory, flying illegally, the single headlight on the front of our little plane shone into nothing, reflecting nothing, giving us no sense of direction. We were completely lost.

No highways were lit up below to aid us. No homes, office buildings, or city streets lit up for the evening. There were no airports, utility stations, towns, or glowing cities. Nothing, but the pitch-colored blanket stretching in all directions around us as far as the eye could see.

Just when we thought we were in as much trouble as possible, a light flickered on in the darkness on the dashboard. And Antonio whispered, "Shit!"

Nothing scared me more at this moment than Antonio freaking out—considering all day, no matter what happened he never seemed to think we were in trouble. When I checked the dashboard and saw the light was for our gas tank, alerting us that we were about to run out of fuel, a thin layer of sweat seeped out of my flesh and enveloped me in its chill.

Until that moment I hadn't even thought about gas, much less running out of it. After all, I wasn't the pilot. I was preoccupied by flying over mountains, turbulence, controlling the plane, falsified documents, landmarks, directions, drugs and other stuff.

Trying to be comforting, Antonio whispered, "Don't worry. We have another tank."

I didn't even know a Cessna had two gas tanks, one in the right wing and one in the left. The light that just flickered on was for the gas tank on the right wing of the plane. Next to it was a gauge for the gas tank in the left wing of the plane. And the left side of its needle was about to melt into the "E" for empty, as well.

And then it dawned on us: We didn't have enough gas to fly back to Truckee or Reno or anywhere—we just had enough to land. We had to find Black Rock Desert and the lights of the festival.

"Aarghh!!!" I screamed. I couldn't help it. How was I ever drawn into this kamikaze flight with this idiot? "You are the worst pilot ever!" I continued screaming. "What's wrong with you? Do you want to die? What kind of a pilot runs out of GAS in the air? What were we going to do after we landed at Burning Man? How were we going to even fly back home, without GAS?" I was hysterical.

"Hey, man, you don't have to yell at me. I'm going to die, too," he said, as if somehow that made things all right.

"Shut up! We don't have enough gas to fly us back to Truckee. We've been flying for 45 minutes since you said it was 20 minutes to go. We HAVE to land this plane. We HAVE to find Burning Man or we're going to die!"

"Okay, I have a plan. We're going to fly really low," he said, springing into action. "Maybe that's why we can't see the campfires."

I was willing to listen to anything.

As we flew down, he turned to me and said, "Your job is to look for any mountain walls and ridges we might be flying into. I'll fly the plane and watch for the campfires."

I strained my eyes looking out of the windshield, but with my heart racing and my head thumping, my eyes started playing tricks on me. Suddenly, all the hairline fractures in the windshield made it difficult to distinguish between moonlight reflecting off the cracks in the windshield glass and the moonlight reflecting off the face of rocks. Every now and then it looked as if we were flying straight into the side of the mountain. I would shout "Watch Out!" and Antonio would veer the plane away from some phantom wall.

We played this deadly game for another fifteen minutes, dipping and turning and circling, no sign of the campfires. And then the light to the second gas tank turned on.

We decided we would land anywhere—Burning Man or no Burning Man. So we looked with our flashlights at the map for a suitable spot. We saw no highways, lakebeds, meadows, or anything flat. We weren't even exactly sure of our location. So, we looked desperately at the ground below, hoping to maybe see a small country road the map didn't bother to show. Eventually, things grew so bad, we even hoped to find the shimmering waters of a lake; anything other than mountain walls.

We flew around desperately for another 10 minutes, like a lost bird trying to find a branch in the middle of the night. But we saw nothing beyond the impenetrable darkness of the Sierras.

With no more answers left and dwindling fumes of gas, we agreed to fly toward Reno.

A strange look briefly passed across our faces. Maybe it was shame for lying. Deep down we both knew we wouldn't make it to Reno. It was at least forty minutes away, and we had no more gas left in the tanks.

We flew in the direction of Reno for another ten minutes, still hoping to cross a road or run into a patch of lit up ground. Twenty minutes had passed since both lights in both gas tanks turned on. I looked at the gas gauge: both needles were passed "E" and pinned to the side of the empty side of the gauge. We were now flying on whatever was left in the pipes leading to the engine.

I looked at the map for the hundredth time, hoping to see something I might have missed before, when I noticed a red box with official looking writing in it. It read: "Instrument Errors in this Area: Electromagnetic interference." I pointed it out to Antonio and the dark circles under his eyes grew darker still.

"I knew something was wrong," he said, sounding defeated for the first time since I met him in Italy. "The compass was turning this way and that," he said, flipping his hand backward and forward imitating the motion of the compass.

I just shook my head, angry at him. Why hadn't he told me? Why hadn't he known about these things?

Antonio looked at the map intently, crumpling most of it in his hand. "I think we've been flying in the wrong direction, but I don't know by how much. I don't know where we are anymore. I just don't know…" He said sounding frightened now. And then he put his head into his hands.

I felt bad for him then, and felt bad for getting so angry with him. I didn't want these angry words to be the last things we shared before we died. So I patted him on the back for encouragement. "It's not your fault," I lied.

He nodded and said, "I know. It's this stupid map." He glared at the map as though somehow it misled us.

I felt extremely alone.

My heart rate slowed, my body began to feel cold from evaporating sweat, and I stared numbly out at the sea of darkness around us, unable to think anymore, no longer able to distinguish the numbing vibration in my head from the droning of the plane's engine. I only waited now for the moment of silence, when the engine would stop and we would plummet to our deaths. I gave up.

As I made peace with myself, knowing we were about to die, a tiny light flickered in the distance ahead. At first I thought this was the portal to heaven opening up for us—the classic white light at the end of the tunnel. I thought maybe we'd already died, the final seconds of our plummet to death unremembered in the afterlife. And then Antonio broke the silence. I knew we weren't dead then, because I was sure he wouldn't be next to me in heaven.

"I think that's an airport light."

What? I thought, and must have uttered some rude comment, preferring the peaceful quiet death over yet another futile, exhausting, misled attempt to live.

"No dude, that's an airport light they put on at night for planes that might need emergency landing. It's an airport beacon," he said.

It was a faint throbbing light in the distance, so tiny it could have been just a miniscule blood vessel that burst on my cornea.

"How do you know?" I asked, wary of letting myself believe him.

"It's flashing white and green, the colors of an airport light," he said, sounding worried. "But it's still far away. Maybe 10 or 20 miles."

I'm color blind, so although I could see the light I couldn't see it flashing green. So again, I was in the unenviable position of having to trust him. Besides having already flown for 20 minutes on empty tanks, I thought this was going to be life's last cruel joke.

We flew like that for five more cardiac minutes, my toes curled up in a ball, expecting the engine to stop any moment. Our engine burned the sweat of our guardian angels, but somehow we made it. As we approached the flashing light, even though there were no other lights on around it, periodically an outline of a runway, a shadow of a plane, and the rippling windsock were visible.

Antonio snapped me out of my disbelief by saying, "Okay I'm going to aim this plane down around where I think the runway is. You have to see where it starts, remember how it's lined up, and guide me to it when I come back around. Okay?"

The plan seemed easy enough after what we'd already gone through.

And so we began our 360-degree flight around the airport. Antonio dipped the plane down, getting lower and lower, shining its little headlight onto the ground. We found the runway, estimated the beginning and end of it. We gauged where we were relative to the runway and other objects that littered this little airport, and then made our way around again, using the objects and landmarks as our only guide to finding the runway in the dark. Tractor…, plane…, cart with a staircase on it…, windsock… and then "RUNWAY! RUNWAY!" we screamed together as the small headlight of the plane reflected off the concrete.

When the wheels of the plane touched down, I enjoyed every bump. We taxied the plane down the runway toward the little office at the end of it, and somewhere in the middle of the runway still fifty yards away from the end, the engine turned off, completely out of gas. The prop stopped dead.

We shouted with joy and hugged each other. We both jumped out of the plane. And, as cheesy as it seems, I lay on the ground on my stomach and pressed against it, giving it a flat 180-degree hug. And then I kissed it.

Antonio wheeled the cart with a staircase on it over to our plane. We climbed up. He opened the gas tank on top of the left wing and shone the flashlight in—completely empty, even the little reserve section. Not a bead of gas anywhere. Bone dry.

"Wow." He shook his head, more fascinated than horrified.

"Wow," I said, more horrified than fascinated.

The airport was tiny; smaller then Truckee Tahoe. Five crop dusting planes and a few tractors sat around the field.

The office was a small wooden structure with three desks. Photos of someone's children and their little drawings pasted on the side of the desks made it seem like a small family run operation. The lights were off inside, a public phone was lit up outside of the offices with three numbers on a laminated sheet under the phone: One for the 24-hour emergency number, one for the police, and one for a taxi service. We dialed the 24-hour emergency number and off in the distance we heard a phone ringing—and realized it was coming from one of the desks inside the dark office. We looked at each other and laughed and laughed. It was hearty, loud, and hysterical laughter, full of joy that rolled out of us; full of relief, anger, and fear, plus any and all bits and pieces of the emotions pent up inside of us from the last four hours. It felt good to laugh again.

We sat down at the picnic table and began to eat ravenously, chewing chunks of salami from the foot-long log, tearing hunks of bread off the bun. We commented how the salami was the best we had ever had and how this damn Italian bread was sooo good. We opened a bottle of wine and passed it back and forth between us, drinking it down like water straight from the bottle, between mouthfuls of salami and bread, letting half of it pour down our chests. Nothing mattered except the food, this moment, and our second chance at life.

After dinner I leaned back on the bench and rubbed my stomach in contentment, preparing to relax for the evening, when Antonio decided he was going to give me another lesson about planes.

He walked over to a plane ten feet away from us and went under its wing. He pushed on a little nipple sticking out from the bottom and liquid poured out.

"You know why they have this?" he asked.

I didn't want to know why. I didn't want to get into a plane again. And I could tell he was trying to lead me toward something.

"No," I said. "I really don't care. I just want to chill and wait for morning. And then we can decide how to get out of here. Besides how come you didn't do that when we took off?"

He ignored me. "They created this spigot so you can press on it and any water that condenses inside these pipes will drain from the bottom of the plane. This way the flow of gas to the engine is uninterrupted by water and allows the plane to fly without any strange breaks from the gas." He finished with a flurry of his hand. He was definitely trying to sell me something.

I didn't have to wait long.

"If you push this spigot long enough, the water clears and then comes gas." He made my eye twitch with the word "gas."

"What are you saying? We're going to steal gas from these planes?" I asked incredulously.

"Yes, you have the great criminal mind." He pointed at me like it was my idea. I didn't smile.

In fact, all I could think about was how this was the difference between the two of us. This was why I could never be a Navy Seal, or rob a bank. Because most ordinary people would call a cab and ride it to the nearest town, then come back and get their plane the next day and fly it back somewhere during the day.

"No, I'm not going to be a thief." I used that as my excuse to keep from getting back into the plane with Evel Knievel.

"Well, how do you plan on getting out of here?" he asked as if this was our only option.

"I don't know, a CAB!" I said condescendingly. "Like most people."

So, we called the cab company, but no one answered. We waited for ten minutes, drinking more wine and eating more food, and then we tried calling them again. We called the cab company for the next two hours and no one ever answered.

Then we called the local police department and told them about our situation. They were neither sympathetic nor helpful. In fact, they laughed and then mocked us and told us to get a cab.

I know. I know. How could I? I'm really not the daring type. The only thing I can say in my defense is the Milgram Experiment.

Let me explain: In the 70s, scientists at Yale University conducted the Milgram Experiment—a psychological experiment to illustrate obedience and denial. Test Subject 1 sat in front of a board with levers that would send electrical shocks to Test Subject 2 in the adjoining room. Test Subject 1 pulls lever after lever, each increasing the amount of electricity to the other person. With each lever pulled, Test Subject 2 screamed in pain, ensuring Subject 1 was well aware of the damage he caused.

When approaching the tenth lever, which had a red line above it signifying it was a deadly amount of voltage, most participants would turn to the scientist wearing a white lab coat and ask if they should continue. The scientist always nodded, and ninety-nine per cent of the participants pulled the lever. Why do we humans behave this way? In order to stop before pulling the tenth lever, Test Subject 1 would not only have to go against the authority of the scientist, but also admit he'd done a bad thing from the beginning by pulling the other nine levers. Interestingly, the ability to correct oneself and stop in the middle of an experiment gone wrong was somehow beyond most humans.

I'd seen footage of this experiment in high school and had always believed somehow I was unique and would inevitably be among the one percent who stopped, regardless of what the scientist said and what I'd previously done. Well, that night in the Sierras, although I didn't think about the Milgram Experiment, in retrospect I realize I'm not so unique after all. I would probably have pulled all the levers.

We decided 40 bottles of our 1.5 liter Crystal Geyser water bottles would give us 60 liters of gas in each wing, for a total of 120 liters, which Antonio said was about the limit of the two tanks of gas. Not knowing exactly where we were, we wanted to over calculate how much gas we needed to reach Reno.

We began stealing gas from the planes in the yard. We would push the mouth of the bottle up onto the spigot and let the gas spray into the bottle. Only problem with the bottle trick is that it slipped sometimes and a fine mist of aviation fuel would come down on our faces and shirts. Before long we were soaked in fuel.

After about an hour passed and we filled the tenth bottle, we realized our task would take at least four hours. Noticing a larger plane twenty feet away, I figured, bigger plane, bigger spigot, faster fuel.

I began emptying gas from that plane and was halfway into filling my bottle when I noticed the color of the liquid was different. Since my color blindness primarily affects the color green, I was able to see that my bottle contained clear, light blue liquid, whereas the first ten bottles I poured into our plane were filled with clear, darker red fuel.

I brought the bottle over to Antonio and showed him the new liquid.

"Hey, where did you find that? That's the gas we need for our plane," he said, nodding his head as though calculating how much of the wrong red gas was already inside our plane.

"Why didn't you tell me? Why were we filling our plane with the wrong stuff?" I asked.

"Well, it's not the *wrong*, liquid. It's just not as good." He was mimicking me. "It's like regular unleaded versus super unleaded."

I was too tired to argue. So together we went back to the larger plane and filled the rest of our tanks with the *better* gas.

It took us over four hours to top off the gas tanks on our plane. As our final act, we went around the airport, leaving a twenty-dollar bill on every plane with a note apologizing for stealing their gas, like some hedge against Karma for this upcoming leg of the flight.

We took off from the little airport into the night, but at about 100 feet off the ground the plane suddenly began to cough, like the engine might sputter and then die. It continued to cough for another ten seconds or so, as though trying to digest something that wasn't good for its system. I thought back to the *wrong*, darker gas we diligently poured into our tanks, but she seemed to burn through it and the plane continued upward and onward.

We flew for about forty minutes in the general direction of Reno, with Antonio reassuring me, "It's a big city, we'll see it when we get near it."

Just when doubt began creeping back in and the old familiar feeling of fear returned, the glow of Reno's lights framed the mountain in front of us, like a meteor that had struck the earth, and washed away our fear.

We taxied into a private plane terminal in Reno and strolled up to the desk with an attractive blonde woman sitting behind it. She casually asked us how our flight had gone, oblivious to the harrowing eight hours we'd just survived. Being back in civilization with the bright lights, slow ordinary pace, and sense of predictability felt surreal. I think I understand the outline of post-traumatic stress syndrome: It's living with the knowledge that the world around us isn't as it seems. What we have is temporal, and just beyond our everyday perceptions is a chaotic hurricane waiting to sweep us away. At least for awhile, until you can forget the experiences of the world beyond, it's hard to relax and enjoy the civilized part of our existence.

We went out that night to a local casino. We talked to everyone, joked with people, and did whatever came to mind. We gambled a bunch and won a bunch—thousands, actually. We ate whatever we desired, flirted with all the girls, and bothered the poor waitresses who couldn't bring the drinks fast enough for us. As if every new interaction was further proof we were alive.

Our eyes met every few minutes and we laughed nervously, disbelieving and guiltily, as if somehow we'd stolen our lives back from the Reaper.

The following morning we asked a few locals where Black Rock Desert was and although they could point it out to us, no one knew exactly where Burning Man was being held. So, exhausted and defeated, we gave up on the idea of landing the plane on the desert floor and decided to fly back to Oakland and try going to Burning Man by car the following day.

It was now the day shift in the terminal. Antonio asked the young guy working behind the counter to fill our plane with gas. He looked up impatiently and quoted us the per gallon price of the gas before ringing it up. Antonio complained that gas was expensive up here in the Sierras and said it would be cheaper to fill it up to only three-quarters of a full tank and buy the rest in Oakland. This made sense, since I knew for certain we had plenty of gas left in our tanks. So the young man filled our order and went back to flirting with the girl.

Strapped into our plane, with three-quarter-full tanks of gas, we began our journey back to Oakland with our tails between our legs. We reminisced and laughed about the night before.

Somehow going over all that horror again seemed cathartic.

We flew back over the Sierras, and it seemed less frightening this time.

We passed Sacramento and could see the bay off in the distance.

This is where the story should have ended—but, not if an Italian is playing a major role in it. In all our relief, somehow Antonio had forgotten something crucial.

Again, only when he whispered, "Oh, shit. But I think we'll be all right," did the words, "here we go again," echo through my head.

I saw his eyes trail off the gas gauges and look intently forward as if he was trying to mislead me into looking out the window and not at the gauges. I looked over anyway, and at first I thought I'd somehow engraved the image of empty gas gauges from the night before onto my eyes, because they both registered empty again. I thought this was impossible, because we'd flown less time than the night before.

"Don't worry," said Antonio, pointing at the bay. "We can make it. We can see it. We're practically there."

But, we weren't. As close as it looked, I realized Sacramento was still almost a hundred miles away from Oakland, and we had no idea how long the gas tanks were empty.

"How did this happen?" I asked, still confused. "Are the gauges broken from what we did the night before?"

"I forgot to thin the gas once we reached altitude. Normally, we lean the fuel mixture so the engine runs smoother and it burns less fuel. We've been burning fuel at maximum rate this entire time. Remember the throttle?" He sounded annoyed, as though it was all my fault because I didn't remind him to lean the gas mixture or remember the throttle.

"The throttle?" At this point I realized he was a threat to humanity. "We have to land at Sacramento."

"No, I'm not going backwards," he said, stubbornly like a child not wanting to admit he was wrong.

Sacramento was a few miles behind us, and I figured by the time we turned around we'd burn more gas. So, I pulled our trusty, gas stained, crumpled map out and began to look for airports between Sacramento and Oakland, hoping to find one just ahead of us. I spotted one that read in big red letters TAFB. "Let's land there," I said, pointing to the letters on the map and out the window, happy to see it was so close.

Antonio looked at the gas gauges pinned on empty and begrudgingly tuned the radio to the frequency of TAFB he read off the map.

He called down to them. "This is November 9-4-1-5, request permission to land." He spoke half heartedly, as though doing me a favor.

"November 9-4-1-5, this is Travis Air Force Base. You are about to enter military airspace. We highly recommend you do not land unless this is an emergency. If not, you must turn now and avoid this airspace. This is a United States Air Force base."

"Well, we're low on gas and were hoping to maybe buy some from you guys," Antonio said, as if it were a logical response.

I was aghast. I imagined that wasn't how we should talk to an uptight military guy who probably had a missile aimed at us. And I doubted he had any gas for us.

"Sir, if you land that plane onto our airfield we will, A) have to cite you for landing on a military landing strip, B) cite you for flying into military airspace and then, C) we will have to take your plane down to the screws, and it will be your responsibility to pack it all up and take it away. Do you understand what I'm saying?"

Antonio looked at me. "Fuck, dude. This plane costs like one hundred thousand dollars. We can't do that."

In a way I agreed with him, but my urge to live was greater. Luckily, the military man made our decision for us, or Antonio and I might have argued ourselves into being shot down.

"There's an airport 10 miles southeast as the crow flies from here called Concord. I highly recommend you land there," he said.

Antonio thanked him and looked at me like, "see, it's all going to work out. You don't need to ruffle your feathers like that. Relax."

We flew over to Concord still unaware how much gas we had left. And as crazy as it may sound, this time around wasn't half as frightening as flying in the pitch-black nothingness of the Sierras.

Finally, we were close enough to hear the Concord tower call to us—a woman with a strong, calming voice. "November 9-4-1-5, you are fifth in line to land. Standby for instructions." Somehow I felt safe within her nurturing, protective grasp.

"Thank you, we'll get in line. What is our approach?" asked Antonio.

"Landing west. Enter left downwind on a forty-five and extend downwind east. I'll call your base turn. You will follow a Citation jet about twelve miles east. Report having the Citation in sight."

Antonio and I flew in a wide circle, then over the airport. South of the field we made a lazy right turn, rolling out toward the runway. All the goodwill between us from the night before had vanished along with the gas.

"Hey, shouldn't we be telling her we NEED to land now?" I asked, wondering what the hell he was up to. Besides, it was taking an awfully long time for our turn to come. We kept flying east until we saw the jet coming in the opposite direction.

Sure enough, he did have something up his sleeve. "I don't want to call an emergency landing, because then she'll write me up. We have to wait for her to call us in." He said all of this as if risking our lives made all the sense in the world.

"9-4-1-5, turn base now, cleared to land after the Citation jet. Caution, wake turbulence. You appear to be very high for our pattern altitude." She gave us clearances and advice in a soothing voice.

Antonio lied and said, "Turning base. Cleared to land. We're practicing a steep approach to landing. Okay?" But he didn't descend, and continued circling at eight thousand feet.

And then he leaned over to me and whispered, "I need to stay high up so if our engine stops we have enough altitude to make it to the airport," as if to comfort me.

On the word "stop," our engine coughed a few times and then ceased running. It seemed to mirror our guardian angel, who was tired and had had enough.

Somehow, you imagine the world in the clouds to be quiet, maybe some wind ruffling over our ears. It never is, except in our dreams. In real life, the engine droning sounds like some comfortable friend, keeping us company and aloft.

So when the engine stopped and all that stared back at us was the propeller standing stubbornly upright, not doing it's job, and the vibrations stopped, leaving a general numbness, and everything was silent except for the wind whipping along the sides of the plane, my heart stopped with it. For about five seconds everything was still. Fate was still unsure what to do with us. Then gravity took over, our plane started to dip forward, and all I could see was the ground rushing up to us.

Antonio turned to me and said, "Oh, God, I fucked up." A phrase I never thought I would hear from him twice on the same journey.

I slapped him in the face, finally, and could only scream, "Just land this fucking plane!"

This seemed to work, because he went into action and began using what little training he had. He raised the nose, slowing us. He pulled on some knobs that pulled our flaps down a little, giving us more loft and steadying so we were no longer pasted to the back of our seats rushing toward I-80.

"November 9-4-1-5, are you having a problem? Did you lose your engine?" The lady in the tower asked with as much incredulity as I felt.

"Affirmative. It did. We request permission for an emergency landing," blurted Antonio—a phrase he should have said ten minutes ago.

"9-4-1-5 you are cleared to land. Do you need fire trucks scrambled?"

"Oh, man, I'm in big fucking trouble." Antonio said. "I'm gonna get written up and lose my license. I'll never be a pilot."

Thank God, I thought.

We glided in for about five minutes. Antonio managed to mention ten times how lucky we were to have stayed at that altitude, so we had gliding room. But the closer we got to the airport, the more it appeared we wouldn't have enough altitude to reach the runway.

A city planner somewhere in Concord should be flogged, for placing a ten-story building directly in front of the main runway. As we approached, and because we were approaching so low, we had to fly around the building and curve back in to line ourselves up again. I even involuntarily lifted my legs, as if somehow to give us more loft. With a few meters to spare, we managed to touch down softly onto the runway.

An ambulance, a fire truck, a gas tanker, a few all-purpose general airport service vehicles, and two security cars all came screeching out to greet us. Antonio and I climbed out of the Cessna surrounded by twenty or so people, like two famous astronauts who were in trouble. A man in a white shirt took Antonio aside and started questioning him. No one seemed too interested in me, so I just stood there taking it all in. Some line service workers hooked the plane up to a small tug and pulled it to the fuel pumps.

As I eavesdropped on the conversation with Antonio, suddenly a man yelled, "You stupid fucking idiot!" All heads turned to the scruffy unshaven man in blue overalls filling up our plane with gas. He stood on a stepladder looking down into our gas tank with a flashlight. I knew exactly how he felt. "Do you know you're supposed to have 30 minutes of flight time left in your tanks when you land? You know you're breaking a law if you don't? You could lose your license, buddy. Who the fuck do you think you are waiting till this late to fill your plane with gas?"

I couldn't help but smile. All heads turned to Antonio waiting for his answer. *Finally,* I thought, *justice.*

"The gas gauge was broken," Antonio answered indignantly. Like somehow it was everyone else's fault the equipment wasn't kept up to date. "Yeah," he continued, "it looked full and everything was fine, and then just after we passed Sacramento, all of a sudden the gas gauge dropped to empty. It was terrifying." And for the Oscar, he then made the sign of the cross and looked up to the sky and said, *"Grazie a Dio,* we were lucky to make it."

I nearly sputtered. I wanted to scream and warn them: "He's doing it again! Don't believe him! How often does a gas gauge not work? He might be flying your kids one day!"

I thought surely they'd see through his lies.

But just like me, they didn't. I watched as his lies soothed all their anger and consternation away like aloe vera on a sunburn.

What was about to be a public hanging, turned into a Hallmark moment.

To my horror, someone even patted him on the back and whispered, "It was a hell of a job you did, flying that baby in."

Antonio was never written up or reported to the FAA. I'm surprised he didn't get a fucking medal.

We got back to Oakland and ran into the old mechanic again when we were about to leave the tarmac. "You guys are back early," he said.

All I could do was nod my head, still stunned, no witty little remarks left in me.

"How was your flight?" he asked, smiling again.

"Up and down," I said, still re-living some of the highlights of the flight, glad to be alive, glad to never get back in a plane with Antonio again, glad to be the one driving the car.

"I told you Charlotte's special," he said and winked. "She never gives up on you."

A little involuntary snicker popped out of me. "No, she's a good one," I said and gave her one last look.

Walking in a Hurricane
by Carol Pilon - Third Strike Wingwalking

There he stood—a tall drink of water if ever there was one, a hand in his front pocket, a cowboy boot exploring the hard pan of the sun baked desert floor and a faint smile playing with the corners of his mouth. His gangly limbs and shy demeanor lent him the air of an insecure adolescent. Only the scuffed cowboy boots and gaudy turquoise adornments spoke for the authenticity of a native desert son. I had come looking for the best. My first impression of the man had left me doubting that I had found him. How could this unassuming individual, too meek for erect posture, be the best barnstormer on the continent? There must have been some mistake.

I had sought him out. I had asked the right questions. I had asked the right people. Time and again, his name was the one that I heard. If he was the best, then he was the one I wanted to learn from. After a year or so of inquiry and several pleading phone calls, he had consented to meet with me. I jumped on a plane bound for Arizona, where he was to be working, looking forward to meeting the best. Now he stood before me, making a complete lie of the preconceived, debonair image I had so painstakingly created for him. For better or worse, I was there and decided to make the most of the visit and learn what I could.

The day passed in awkward conversation, meaningless trivialities, and in closing, an invitation to see him fly at a later date, in another place. We held palaver, but I had yet to see him fly.

* * * *

I had slept on the hangar floor the night before the air show. I was at the appointed place on the appointed time, some 100 miles outside of Wichita, Kansas. There were no hotel rooms available anywhere, and I was limited to a taxi for transportation as I was too young to rent a car in the great state of Kansas.

The lack of traveler amenities didn't concern me in the slightest, nor did they dampen my spirits. I had been invited to see the best, and I'd come to see the best. I was there, but the flying legend was nowhere to be found, and all of my messages went unanswered.

Eventually, he appeared. Somewhat to my dismay, he acted as though he had no idea what I was doing there. Just what our fledgling relationship needed—more awkwardness! At this point, my opinion of him was growing quite dim. I can only imagine that he was more than likely hurriedly implementing anti-stalker counter attack measures.

Showtime was at hand. Being shunned and snubbed by barnstormer and company, I put my time to good use by begging a flight back to Wichita with some war birds. I had also divested myself of the camping equipment I had purchased the night before. A sleeping bag, a portable Bar-B-Q, and a cooler had all found new homes. There was nothing left to do but wait until the show was over so I could leave. Good riddance to a wasted trip. I found a nice patch of grass, removed from the air show staff who I suspect were beginning to think me a penniless vagrant, and resigned myself for the duration.

My neck snapped to show left as I heard the takeoff roll. The sound instantly evoked visions of grand barnstormers of eras past. And there he was, rolling on takeoff a mere 300 feet in front of me with a wingspan of altitude to his name and a streak of smoke chasing him down the field. The next 15 minutes held me spellbound. My eyes never left the aircraft and the whole time, I wanted to be up there with him. From Cuban through Lomecevack, from hammer head to knife edge, from barrel roll through the flat inverted with the engine knocking out and flames shooting from the stack, I wanted to be up there with him. God, how I wanted to be up there with him! Finally, I had met the man that I had been seeking. I had met the best barnstormer that my era was likely to produce, and he'd blown me away. This man was to become the beginning of my life as a wingwalker. He, among all others, left the most indelible impression on my life.

That day, the man became my hero. Later, he became my friend.

I doggedly pestered him every other month for his mentorship. After seven years went by, and I presented him with proof that I could be a worthy student, courtesy of a trial wingwalk from Silver Wings Wingwalking, he became my mentor and we flew.

He became my lover and we flew. He became my husband and on our wedding day, we flew. A short 18 months later, he became my ex-husband and we flew without each other, leaving me to wonder if he felt the same empty space in his wake that I felt in mine. He died and I fly on. His name was Jimmy Franklin, and there will never be another like him.

Despite our trying, blissful, difficult, passionate, and downright ornery relationship, he bequeathed me the greatest gift to which one such as I could aspire. He had made me a wingwalker. He had in fact, made me a solid one. I fly because of him, I fly in spite of him and always, I fly in honor of him.

During my time with the barnstormer, I was introduced to an inspiring woman by the name of Debbie Gary. She was and still is an aviation legend in her own right. I liked her from the first time I laid eyes on her. She possessed a love for life that was lustful beyond description. Her optimism was contagious and her energy, infectious.

She stumbled out of her Cub with a bounce in her step. I had never met a more beautiful spirit. Debbie had also given me an equally precious gift. It was proposed over libations, as truly great ideas often are, that we should form the core of an all-woman wingwalking team. Up until this point in my career, I had represented the unknown, radical element in all of my flying equations. Now, it would be my turn to trust an unproven, although highly qualified, candidate with my very life. This little exercise allowed me to fully understand the enormous risks that both Silver Wings and Franklin had taken on my behalf. The revelation was humbling, to say the least. Debbie and I executed training, developed a routine, and flew it for spectators in Missouri all under the guidance of our now shared mentor. On airshow day, I looked down in horror at her scantily sandaled feet. Taking stock of the entirely inappropriate footwear for an aerobatic routine, I exclaimed, "God, I hope you're not planning on wearing those for the flight."

Her response was the perfect white lie. "Of course not, my shoes are in the trunk." And the show went on with a wingwalker in ignorant bliss of the offending flip-flops dancing on the rudder pedals. What Debbie gave me was a whole new confidence in myself. I had no idea that such a thing even existed within me. Although my time as a pupil was far from done, I had now become a teacher, capable of passing on my knowledge to others and capable of trusting in others. Her gift was that of self-empowerment and self-knowledge. It was nothing short of a thundering awakening, and her gifts would be drawn on time and time again.

Pursuant to my separation from Franklin, I admit that I spent a great amount of time angry: Angry at myself for my obvious matrimonial failure, angry in my misconception that I had suffered unjust loss, angry at others for perceived slights. I was mad, mad, and holy cow, mad. It was a blessing in disguise, for without this rage seething inside me, I never would have had the tenacity or fortitude to achieve what was to come next. It was to be the most challenging stunt that I had yet to orchestrate and survive. This time there would be no teacher to hold my hand. There would be no advice or mentoring to be sought. There would be no precedence to follow. It would be me and my grudge, bound for the heights of success or the dismal pits of failure. I marched into my bank armed with an elaborate yet wholly inaccurate business plan and secured the loan which would allow me to purchase my aircraft, the base of my new empire. Without fanfare or recognition, I had just formed the first wingwalker-led and woman-led team in North America. I had my aircraft, I had a plan, and I was ready to conquer the world. Huzzah!

I went into the venture, nostrils flaring and hell bent on showing the world that I had no intention of rolling over like a kicked dog and playing dead. All of my previous years of dedication and service to the art of wingwalking would not be so cavalierly banished simply for the want of an aircraft, a business, and a pilot. I would stand for no such travesty. In hindsight, I really have to laugh at my complete naivety and self-involvement. The real business of becoming the first woman and first wingwalker to establish a team in North America was at hand, and I needed to answer the call. My anger was swiftly swept aside by financial difficulties, securing competent pilots, convincing air shows to hire a wingwalker-led team, overcoming industry gossip, and dealing with aircraft difficulties that seemed never-ending. Every time I turned around, there stood another bleeding government entity politely informing me of a rule or regulation that I was in jeopardy of contravening, ranging from immigration to taxes to the department of motor carriers. When I'm asked how I stay in shape for wingwalking, I answer that I bench press my paperwork—and it's not entirely untrue.

Having bitten off more than I could chew, I did the only thing that I could. I started to ask for help and I asked often. I used to be fiercely independent. Those days are gone. I mourned the loss of my former self-reliance and begrudgingly embraced my new humble and needful demeanor. Had it not been for the timely intervention of several helping hands, I surely would have succumbed to the multiple burdens that threatened my fledgling survival. These human beings restored my faith in humanity. They banished the last wispy threads of anger from my life. They have been better to me than I deserve, and I love them all for making my dream a reality. I need to share some of their stories because their stories are the ones that molded the person I became.

While I was performing at an air show, a gentleman who was known to me came up to the fence to say hello. He knew that I was a wingwalker but wasn't really moved by the fact until he saw the performance. He told me that my airplane was an absolute beauty, told me that I was absolutely nuts, and wrote me a check for a thousand dollars, just to help me out. I almost fell over in amazement and shock. I was to find out later that his act of generosity was not singular. The list of those who have helped me along my precarious path only seems to grow with time.

Asking for help also led me to Bluestem Aerial Sprayers. Once a year, I have the distinct privilege of dragging my aircraft to Kevin Brown's base of operations in Cushing, Oklahoma, where I am fortunate enough to have the air show princess snot kicked right out of me as I get handed the very real facts of life pertaining to my meager mechanical skills. They are a group of hard-core sprayer pilots and mechanics who have undertaken the very daunting tasks of both my aircraft maintenance and my aircraft maintenance education. They have accepted the heavy burden of maintaining an aircraft whose sole intent is to perform in the highly demanding airshow arena. This burden comes with the knowledge that there is no room for error. The craft is required to output optimum performance at all times. This type of maintenance requires no less than excellence. I trust them to secure my aircraft maintenance and they, in turn, trust me to perform safely.

I can't even begin to explain what this place is like. Once I arrive, I am no longer the acclaimed airshow heroine. In that place, I'm merely the hangar grunt, and I accept all the humiliation and debasement linked to the position. Of course, this is somewhat exaggerated. During my stay, several will stop by the hangar to visit with the wingwalker of some small repute. This practice is still somewhat acceptable to the management. With about five mechanics involved, the debates that ensue every time a mechanical issue arises are vocal, discordant, and bordering on physical aggression. The guy who ends up being proven right continues said vocalization for years to come, as is his right by the tribal laws of the hangar. I pay them in Canadian beer and wingwalking DVDs because they are kind enough to point out that I could never aspire to afford such prestigious and sought-after maintenance services in my humble lifetime and bid me not to insult their talent with what meager funds I could appropriate. They help me because they want to see me succeed, but mostly because they love me.

We have discussions pertaining to world politics in which I am promptly outnumbered. Being the token Yankee amid the rednecks does nothing to help me once we enter into the political arena. When I get uppity, I'm often referred to as their foreign aid case. We have daily breakfasts at the local café, where I try my best to avoid the gravy. A full-blown cookout ensues at lunchtime in the hangar with various wives, tolerant of the female wingwalker in the midst, contributing home cooked fare, or with the appointed chef grilling up some brats and burgers. We war over the benefits and side effects of muffins and gravy. We poke fun at each other at every opportunity, and they are always ready, regardless of my feminist tendencies, to find me a "nice husband" to settle down with. Thus far, most of the candidates have been rejected for want of teeth. I'm not really sure how we ended up being so close, given the disparity between our backgrounds, but I, for one, consider myself quite fortunate to call them my friends. I would be a lost cause without their support. If my life could be judged by the company I keep, my association with these people would lead me straight to the Pearly Gates. They even take time out, when I'm a little blue, to lift my spirits.

When I told Kevin that I was writing about him and advised him that he may want to pursue a defamation of character suit, he bluntly responded, "There aren't enough words in the English language to adequately defame the characters that hang out here." My only complaint is that sometimes, the dogs can smell real bad, and it's usually then that they seek out my lap. Dogs carry weight in Oklahoma and tolerance of olfactory insult is absolutely mandatory. Amidst all our giggles and fits, we manage to get the job done.

On a lighter note, when I masterminded the team, I was adamant that I would take up as much responsibility as possible, leaving only the flying to the pilots. This meant that I would be ground hauling the aircraft. I designed a trailer in which to stow the aircraft and a rig which would help me assemble it. Everything would be driven to the event site by yours truly. I achieved full federal motor carrier status in order to do so. I also somehow managed to get an airframe technician certificate along the way so that I could assemble the aircraft myself, and Bluestem has gone a long way to help me expand on that knowledge. Once, when I was pulled up at the diesel tanks and fueling my truck, there was a cowboy over in the next lane who was scrutinizing my rig. He was hauling a horse trailer on the back of his truck. He told me that I had a nice outfit and asked what I was hauling. I answered him that it was an aircraft. He gave me a complete look of consternation and said, "Why in *tarnation* would you drive an airplane anywhere?"

I looked up into his eyes, nodded my head towards his trailer, and responded "Same reason you drive your horse around, sir." He looked at his trailer, and I could see the realization of the situation dawning on him. With a, "Well, I'll be …" he tipped his hat, bid me a good day, and climbed back into his truck in obvious embarrassment. As long as I live, I will never forget that cowboy. He struck a chord with me. He was all discomfited and mortified. If simply ground hauling an aircraft stymied him that much, I can't help but wonder what his reaction would have been to the whole wingwalking deal!

I have been helped by many individuals in many different ways. I've been offered free welding services and on-the-house maintenance at several different events. Last year at CFB Bagotville, I had at least half of the line personnel congregated around my craft trying to fix a problem with the smoke system. They were all wearing smiles but given the circumstances, I was the token moody performer of the day. Once, as I was walking down a show line with my benefactor, a bucket of Kentucky Fried Chicken was thrust into my hands with the person urging me to eat more chicken because I looked too skinny to be a wingwalker. Another time, someone gave me a laptop carrying case after he saw me wrap up my PC in bubble wrap. A gentleman even sent me fifty dollars in the mail to help put a couple of liters of gas in the tank for practice.

A wonderful woman knocked on my trailer door one morning. She had a little parcel all wrapped up and offered it to me. I was amazed to find a glorious little feather-adorned vintage hat under the wrapping. She had heard that I was a collector of vintage hats and offered me one from her personal collection. How in the world she knew that I was a collector was beyond me.

All of these various acts of kindness and support led me to understand one thing. These people have taken the time to know me. These people have a vested interest in my success, and I owe them my very best. I strive every day to be equal to the faith, confidence, and kindness they have shown me. Wingwalking is not wholly about stunts or airplanes or success. Wingwalking is about being responsible for the trust others have invested in me.

Now let's talk about the trust I place in others: my pilots. First, I will take a moment to familiarize you with Wingwalking 101. Wingwalking is about hanging on, holding on, grasping to, clutching onto, and did I mention hanging on? Once you understand this basic, underlying thread, you can begin to incorporate the rest. There are two sections to our routine: the aerobatic sequence and the acrobatic sequence. During the aerobatic sequence, my pilot is maneuvering the aircraft through different aerobatic maneuvers including eight point loops, hammer heads, barrel rolls, aileron rolls, Cubans, and other flip-flop stuff. During aerobatics, I, as a wingwalker, am required to be in one of two lockdown positions. The first is in a prone position on the javelin between the upper and lower wing, where I hang onto the flying and landing wires for security. The second is secured in the rack on the upper wing where I use a restraint harness. The end result is that I am secured and the pilot can focus on executing maneuvers without having to worry about me.

During the acrobatic segment, I reverse roles with the pilot so that I am now doing the stunts and the aircraft is maintaining level flight. This allows me a stable platform on which to walk, execute headstands, and the rest of the death-defying stunts in our repertoire.

I have found over the years that while a great many enjoy viewing wingwalking, precious few understand what is actually involved. This phenomenon is not exclusively the domain of the spectator. Many professionals within the airshow industry are also in the dark in regards to the art. Much pomp and circumstance has been utilized over the years to make it appear effortless. Trust me when I tell you that it is anything but. From the ground it all appears so fluid and graceful, but up there, in my world, things are about to get intense. During aerobatics, I'm exposed to winds up to 160 miles per hour.

Now, consider that a Category Five hurricane will develop sustained winds up to 155 miles per hour. This is when buildings start to come apart. Luckily for me, the only debris that I need to worry about consists of bugs and birds. Bug strikes are pretty painful. Suffering a bug strike causes what is know as a pain reflex. The emergent consequence of a pain reflex is to protect the wounded area. This also means that you have unwittingly let go of the aircraft. Yikes! In my world, that is unacceptable. There have even been some cases of penetration. Yes, the bug actually penetrates the wingwalker's uniform and skin. We call that an "owie." So far, I have avoided this unpleasant scenario. Suffice it to say that a bird strike would indeed be a very bad day at the office. Thankfully, birds try to avoid people strikes and give us a wide berth.

A normal routine has a G-load range of 0-5+. For the uninitiated, a G-load is the gravitational force acting upon your body during aerobatic maneuvers. The easiest way to describe it would be to imagine that your body weighs five times its normal weight at the high end of the scale, and that you would weigh nothing at 0 G-load. But that's not the half of it. When you start applying G-loads to your body, all of your blood has a nasty tendency to want to migrate to your toes. If you fail to prevent your blood from leaving your head, you will have an impromptu nap at a most unfortunate time. Going to sleep on the job when you're a wingwalker is just plain bad for business. The only way to combat the effects of gravitational force is through a series of timed muscles contractions, repeated every time the nose of the aircraft points down. G-load also means that sometimes I can't move. Excessive winds hinder my breathing. The inability to breathe freely can lead to lactic acid buildup in muscles, causing them to be inefficient. I have to time my breathing like a swimmer in order to be effective. Continued exposure to the cold caused by the winds also hinders muscles. This renders movement sluggish when speed of movement is required. The wind velocity, now coupled with the physical exertion of staying on the wing, sees hydration levels dramatically taxed. Lowered hydration levels can help the G-monster put you to sleep faster. Contusions ranging from mild to "someone beat you with a two-by-four" are commonplace. Bruising diminishes in occurrence commensurate with a team's experience level, but one gusty day and you're sent, limping, back to the drawing board. So I guess that wingwalking is really like trying to stand on your head on a roller coaster ride in a hurricane for 15 minutes while fighting off insomnia with a gremlin trying to push you off. Oh, I forgot the most important part: Our team needs to make it look easy. Pft! No problem.

Alrighty, then! How do you prepare a pilot for the wingwalking nonsense? First, you must locate a willing victim with the proper credentials. Then, you ascertain if the benefits of the union will outweigh the risks. If everything looks good, you begin by talking. Once I'm satisfied with the pilot, I boldly hand over my beautiful aircraft to him or her, proceed to the nearest restroom and vomit at will while the new pilot makes the first flight. Once I appreciate that the novice pilot can land the aircraft, anxiety leaves, and I can watch her gradually progress through the routine until it's time for me to join her.

I have discovered a little point of interest while training new pilots. The first time they see me get out of the aircraft and climb into the upper wing rack, they get scared, and I mean really scared. I turn around to signal them only to find them wide-eyed, drop-jawed, and ashen white. Handing control of my aircraft to a new pilot terrifies me at first. My terror, however, does not begin to match what they experience the first time I start walking around while they're flying. I now include this expected "state of shock" in our pre-flight briefs. They are mentally prepared and briefed on the sequence of events, but the first time I get out there, it freaks them right out. They see it happening but they experience difficulty mentally computing the fact.

You need to bear in mind that these people are going against years of seatbelt indoctrination. They can't even abide a loose strap. They seem to get over the initial shock fast enough once they remember that they need to fly the aircraft. From this point on, we just keep adding various maneuvers and wingwalking segments until the routine is complete. Once satisfied with the routine, we bring in an ACE to review, assess, and accredit the pilot with wingwalking certification. An ACE is an Aerobatic Competency Evaluator, the appointed official who will determine if we are competent to perform our routine on the airshow circuit.

Once, I had trained a pilot only to find him missing in action when I required his services. Some months before our first scheduled event, I couldn't locate him. It was an embarrassing situation, to say the least. There I was, getting ready for my first air show with my new team, and I had misplaced my pilot. I called my ACE, Kirk Wicker, to fill in. Not only did he save the day, he also ended up spending the better part of two years as lead pilot for the team.

While together, I received a call from a gentleman inquiring about our services for an event to be held on February 26 and 27. I thought everything was great until he disclosed the location. It was to be in Maine, during an ice-fishing derby. I could not believe what he was asking me. Didn't he know that it couldn't be done? Didn't he understand that I would freeze to death? Even though I had thought him quite insane, I kept listening to him talk. Before long, he had won me over to his cause.

I called up Kirk and asked him what he thought about it. It was kind of funny listening to him run down the same list of impossibilities that I had just stated to the show producer. After some lengthy discussion, we came to the conclusion that not only could this feat be accomplished, but that we, in fact, would accomplish it. The producer was ecstatic to hear our acceptance.

To the best of my knowledge, no one had ever attempted a full-out winter wingwalk before. It was going to present some unique challenges. Most likely, anything exposed to these temperatures would fall off. I'm rather fond of my various body parts, so resolving this first issue was of paramount importance to me. We would have to deal with several unknown factors also. We would have to implement countermeasures against all forms of nasty side effects related to the cold. Kirk would even have to figure out how to land on the ice with wheels. After Kirk and I conferred at great length, we devised a plan of attack, taking all into consideration and attempted the first-ever documented winter wingwalk together.

Everything went great and it was a resounding success. We repeated the same feat the following year. This time we didn't have the benefit of performing from the lake surface as the water wasn't frozen. This required departure from a remote site and entailed 45 minutes of yours truly aloft in some seriously inclement weather. Unfortunately, this allowed us to figure out my limitations in regards to exposure. I had never been so cold in my life. It was beyond painful. It was beyond excruciating. Absolutely every body part hurt, including my teeth. Once, as a child, I had fallen through frozen ice. This experience was worse. After the show, I did manage to find enough hot toddies to allow me the semblance of forgetfulness.

We had invited the producers to dinner after the show. They brought their families. Rob Holland was also there, as he was flying the show in his own craft, and we were joined by one of the sponsors and the crew who came down to help me assemble the Stearman. It was one of the most pleasant evenings that I can remember. We were all gathered around the table enjoying each other's company. The jokes at my expense that evening were abundant. We had fought the good fight and lived to tell the tale. Our respective battles with the cold had somehow brought us all closer to each other. And in case you are wondering, I would absolutely, no holds barred, do it again in a heartbeat—not because I like the cold but because I have never in my life found a more appreciative or receptive audience than the one that was there to greet us when we landed.

From the very beginning, Kirk was the strongest supporter of my dream. His early assistance and his unwavering confidence in me inspired me to continue when things seemed bleak. When I faltered and when I failed, Kirk remained my most adamant supporter. He picked me up by my bootstraps, dusted me off, and set me back on the path. Once again, I was reminded of the trust that others had placed in me, and of my duty to uphold and preserve it.

The first year of the winter wingwalk, Kirk and I watched a young man by the name of Rob Holland flying a hard deck of 250 feet in a Pitts. Kirk strongly suggested that I invite Rob to try out for backup pilot position on the team. Rob accepted the invitation and (as of this writing) has now been lead pilot for the last two years. Training with him was completely effortless. He kept asking how he could improve, and I kept saying, "You know that thing you did earlier? Do it again, the same way." I have rarely seen such a natural born stick. With only ten hours on type, he took to the aircraft like a fish to water. Now that he has progressed through to the newest type of aerobatic aircraft available, he finds plenty to complain about with the Stearman. Although the Stearman is a grand and beautiful craft, it's nothing more than a drag-making machine compared to the sleek and performance oriented crafts that Rob is accustomed to flying. He came with all kinds of new routine ideas, and he quickly implemented them. In so doing, he stepped up the aerobatic portion of the routine substantially. A good friend had once told me, as only a good friend could, that my days of flying with the likes of Jimmy Franklin were done. I would never find another pilot anywhere like him. I should stop searching, because that illusive pilot simply did not exist. My friend was, of course, right, but he was also somewhat wrong. Rob Holland fills the cockpit as well, if differently, than ever my barnstorming mentor did. I see many of the qualities that drew me to my mentor reflected in Rob's flying. Comparing them would not do either justice, but suffice it to say that I have found a harmony with Rob that is as near perfection as I could hope to achieve. We fly together and we fly apart, but it is always a joy to turn around and see him once again sitting in sweet Rhapsody's cockpit. The best thing about Rob, however, has nothing to do with flying. I tend to be a little over anxious because I'm one of those high-strung types. Rob, on the other hand, is solidly consistent. In direct conflict to his maniacal flying style, he possesses the most grounding human qualities. He keeps me sane. He is the voice of reason. I defer to him on a great number of things because I value his input and opinion. No matter what I throw at him, I can be assured of one thing—an absolutely level response. I truly hope that Rob will be in my life for a long time to come.

As I always seek my pilot's guidance in these matters, Rob quickly suggested Todd Schaufenbuel as our new backup pilot. Funnily enough, this happened at yet another bar. Todd had been flying with the fabled Red Barons but a hiring freeze within the corporation prevented them from renewing his contract, and I was in the market for another backup pilot.

Todd was in for a treat because Kevin let us use his base of operations in Oklahoma for training. It wasn't long before the place was filled up with pickup trucks and dogs. The guys even let him fly one of their spraying aircraft, which Todd promptly used to terrorize every stem ragweed in town. Todd will now be in his second season with the team as backup pilot. Todd knows Stearmans. He understands the aircraft inside and out, and he doesn't mind getting his hands dirty. He manages to extract all kinds of energy out of her, and he doesn't let her get away with anything. He is still a little leery of me as a boss, though, so he lets me get away with almost everything. Todd is accepting of any new idea or stunt that I throw at him. He is open to any and all misdemeanors, and I love him for it.

All of the pilots I work with have completely different styles of flight. Their presentations, along with their sequences, are unique unto themselves. I have found that a pilot can be a most compelling performer when he is given free rein to extract the best from both the aircraft and the sequence alike. Pilots are an inclusive part of the team, and I value what each brings to the team. They let me wingwalk, and I let them pilot. So far, this state of affairs has seemed to work well for all involved.

I have often been questioned about the relationships between myself and my pilots. A typical team includes a pilot, who is usually the owner of the team, and his wingwalker, who performs on a contracted basis. Establishing a wingwalker-led team now meant that a roster of several different pilots would alternate while the wingwalker would remain the constant element. There I was, rewriting the rules to suit my fancy yet again. The formerly held team hierarchy fostered the long-held myth that both pilot and wingwalker exercised an almost telepathic bond, which allowed for minimal communication during the routine. While I would love to boast telepathic abilities, no such thing exists in aviation; if it did, only fools would rely on it. Much like formation flying, any team member, at any given time, can be replaced by a candidate of equal skill level. The same holds true for members of a wingwalking team, including myself. Our communication is indeed quite limited, but we do not overcome this issue by means of sixth sense. It is overcome by training, consistency, and practice. While I do enjoy very close working relationships with all of my pilots, they are not founded on mystical flimflam. They are based on a profound respect of each other's skills, talents, and the confidence born from repeated, successful sorties.

When asked about the inherent dangers of wingwalking, I return to training. Throughout history, training and dedication have always been the keystones to successful wingwalking routines. Wingwalkers are no longer the insane stunt persons who pervaded the early days of barnstorming.

Our team consists of trained, talented, and serious professionals, using highly modified, advanced, and exquisitely maintained aircraft, with the best of pilot skills protecting me through every step of the routine. The most important part of my routine is the pilot. Each pilot has the very real responsibility of not only executing the routine but of also protecting me during the flight.

Although I am the boss of this outfit, the pilots are incontestably in command once the wheels leave the tarmac. This means that they, in fact, control every single movement which I execute. I never seek to move without pilot approval. Everything you see happening during a wingwalking routine is courtesy of an excellent pilot setting up the optimum conditions to allow me to safely execute my role.

Once, a newspaper article erroneously reported that I controlled the flight by stomping on the wings. These offending lines earned me a weeks worth of dark, malicious pilot glares. I do not seek to understand or comprehend commands given by the pilot. I simply act upon the command. Some would call this blind faith, but it is nothing of the sort.

By the time a pilot is performing at an event, sitting in my cockpit and telling me what to do, you can rest assured that he has earned my trust, and trust him I shall. He has spent several hours of training in which I have been suspicious and questioning of his every command. I have tried his limits and come to understand his particular quirks. At show time, we each need to maintain our respective roles to insure safety and proper execution. Everyone on the team must maintain his or her appointed position. Failing to do so could only result in chaos. My role is to wingwalk, theirs, to aviate.

Letting go of responsibility is a very hard thing to do, but it is a necessary element of a successful routine. I hand the pilots the mantle of trust, and I expect them to be equal to that trust. These individuals have spent the better part of their lives honing their skills and talents. They have each faced peril and survived. I have no problem following their lead.

The most frightening scenario that I've encountered as a wingwalker was particularly dangerous. The danger was so insidious and deceitful that I barely managed to see it coming. While this description sounds suspiciously like the G-lock onset which I once suffered in the midst of a routine, it is far more dangerous and debilitating. The biggest threat that a wingwalker can ever face resides exclusively between her two ears. As any airshow performer can tell you, we get plenty of acclaim and glory pursuant to our respective performances.

As a wingwalker, you get a truckload more. Everyone thinks that you are fantastic regardless of whether your performance was stellar or lackluster. They all think you are positively fabulous and can't wait to tell you so. The papers want to picture you, the television wants to interview you, the young girls want to be you, and the guys are intimidated by your fearsome prowess. They all just love you to death. It's all so very intoxicating and alluring. The temptation of falling prey to one's own propaganda is a very real threat.

While we all like a little recognition and appreciation in our lives, this kind of attention, if not held in check, can end up turning one into the dreaded air show diva! I have seen a good number of victims to this scourge during my time. When the "I" welfare replaces the team welfare, you might as well pack up and hang your shoes on a peg, as your situation can only lead to a bad place.

My guard against the diva phenomenon can be found in a small group of my air show peers. They are the only ones I will listen to in regards to how good or bad a routine may have been. They tell me the absolute, candid truth about where I stand. In exchange for their brave honesty, I have consented to slamming the door on the flood of seductive whispers. I enjoy nothing more than meeting with the spectators. I thank them for their compliments and support, but I don't believe a single solitary word of acclaim as it relates to the execution of my wingwalking routine.

Our performance may well have been earthmoving to some in the audience, but to our team, it must remain merely the execution of a successful routine. The attention is pervasive and insidious. It can affect your judgment of yourself and your limitations. It can negatively impact team dynamics. It can lead ultimately to your own demise. Playing the part of the airshow star is very important. Stars and heroes are what the audience has come to see, but it is merely a role played as part of our services. If one truly believes that she is a star, she needs to be in Hollywood, not on a wing.

When you look at your pilot and give the head nod, signaling that you are ready to start, you had better be authentic, dedicated, and nothing but professional right down to your toenails.

My whole life, from the very first time I witnessed wingwalking, has been about becoming or remaining a wingwalker. I knew from the word go that this is exactly what I was suppose to be doing with my life. No reserves, no questions; this was my path. It took me seven years of pleading and begging before I ever stepped out onto the wing of a moving aircraft. That was the real school of hard knocks. Seven years of rejection is a bitter pill to swallow.

In revenge, my training and schooling were a whirlwind of activity. I had learned, in the span of a mere two weeks, from the best of teachers, what it would have taken me a lifetime to learn on my own. Two weeks and I was wingwalking at an air show. I couldn't help but remember my trip to Kansas. I was there; I was up there with him! Wingwalking saw me become the first and as of yet only woman to walk on a jet-propelled aircraft. It saw me working with multiple pilots, now numbering 11. It saw me train to accredit no less than four pilots. It has seen me walk on five different types of aircraft, including the most powerful at an output of 2,000 horsepower, to the most underpowered at 65 horsepower.

Wingwalking has seen me execute the first and second documented winter wingwalks. It saw me through formation walks, nighttime pyro walks, dual walks, and solo walks. It has seen me train other teams, entertain audiences, and educate youth. Wingwalking has seen me become the first Canadian wingwalker in history, and it held my hand when I formed the first wingwalker-led and woman-led team in North America.

Wingwalking has led me to many benchmark accomplishments, despite the fact that none of them was ever the goal. The goal has always been to be a wingwalker. Everything else was simply a means to an end. Some day, with perseverance and hard work, I may just become one yet!

The question remains: Why be a wingwalker? The simplest answer is the truest answer: Because I must. There exists only one true form of freedom: The freedom to exercise your option of responsibility. Wingwalking is both freedom and responsibility. The answer can also be found in the feeling I get while clinging to the flying wires in inverted flight. In that moment, I am Superman. It can be found in a dizzying heels-over-head tumble when the horizon slips under my body, and I am magically suspended above the earth. It also lies within the chill of a vertical dive when the crush of gravity is pulling me inexorably closer to earth. The answer can be found in the complete and total abandoning trust that I place in my teammates. It can be found in the eyes of an inspired child looking up at you. For the fifteen minutes that the flight lasts, I belong entirely to the moment at hand. Nothing else matters save the immediate and my reaction to it. It is both exquisitely liberating and painstakingly precise.

Once you have accepted the responsibility of upholding the trust that others have placed in you, there is simply no denying it and no turning back. I am a wingwalker because somewhere between throttle-up and landing, I am given the gift of ultimate freedom. How could anyone refuse such a thing?

Throttle it up, Baby!

The Wingwalker—

My life is hard, my story old,
But I have dreams and they'll be told.
I'm wingwalker born, wingwalker true.
What price for all that I've lived through?
Friends and lovers dead and gone,
Yet I remain, to carry on.

Your widow's cries and wails I hear
But for you, my friends, I shed no tear.
We can reap only what we've sown.
The only life I can save is my own.
I'm wingwalker born, wingwalker true,
But my heart is hurt, and I miss you.

Rhapsody simply yearns to fly,
To tear up every inch of sky.
Her lessons hard, her heart so cold,
To tread on her, one must be bold.
Her pulsing engine cry I hear.
She calls to me and I know no fear.

I will hear the wind howl out its rage
As we break through the dawn-lit haze.
Its bite will burn and singe my skin
But there's no way I'll let it win.
Toss and tumble, fall and kick,
She won't shake loose my iron grip.

I will dance upon my balsam steed
Across the sky at breakneck speed,
So glimpse me if you think you can
While I tempt fate with open hand.
Fear what you will but not for me,
For this be my love, my destiny.
I'm wingwalker born, wingwalker true,
And the things I do, I do for you.

I stand before you unashamed
And to my birthright, I lay claim.
When gauntlet thrown, I answered call.
I've laid down my life before you all.
Don't dare to judge me from on high.
Heed my rebel cry, I'm alive so let me fly.

~~Carol Pilon

Carol standing by her Stearman, Rhapsody, before an air show.
Photo courtesy of Chuck Hodgdon

Dancing on the Stearman as it does a high-speed, low pass during an air show.
Photo courtesy of Eric Dumigan

Wingwalking while Stearman does barrel roll. Carol is inverted in this photo, her head toward the ground.

Photo courtesy of Eric Dumigan

Carol Pilon hanging from wires while pilot gives "thumbs up."

Photo courtesy of Eric Dumigan

CHASING PABLO

by Nick Qualantone

Hippos. The largest herd of hippopotami outside of Africa resides near the small village of Puerto Triunfo, Colombia. A local hotel operator is refurbishing *Hacienda Napoles*, the 22-square-kilometer former home of the hippo herd, formerly owned by cocaine king Pablo Escobar. Concrete dinosaurs built by the drug lord as part of his fantasy theme park are being fixed and painted; the luxury prison Escobar built for himself is now a tourist attraction. People come to see the burned wrecks of Pablo's classic car collection, torched by vandals.

Mostly they come to see the place where the world's most notorious drug lord lived. To many, Pablo Escobar was a folk hero—a modern-day Robin Hood who built football pitches, parks, and churches, and provided jobs for local peasants to keep their sympathy and support. To the governments of Colombia and the United States, Pablo was a murderer, responsible for importing huge quantities of cocaine into North America during the 1980s.

During the last year of Escobar's life, I became part of a massive search to bring him down, along with the Medellin Cartel.

I chased Pablo.

In November 1993, I worked as a pilot for a Special Operations unit headquartered near Washington D.C. We were a highly secret, signal-intelligence gathering organization. In fact, we operated under such deep cover that only the President and his closest cabinet members knew we existed. Our carefully constructed cover was thorough. We wore no uniforms, carried no identification, and each of us had a plausible, totally false personal background we had memorized, tested, and retested. We operated under *alias cover,* with each of us holding *operative status* assigned and supported by the CIA and the State Department. Our missions were approved at the highest level, usually by the President himself.

Operation Centre Spike was one specific operation, now declassified. Centre Spike actually dates back to 1989, when Americans secretly flew Beechcraft 300 and 350 aircraft, the King Air and Super King Air, and tracked Pablo Escobar around the country with special electronic surveillance equipment.

This operation was only a part of the United States' anti-drug trafficking effort. In the late 1980s, the Colombian government requested help from the United States to gain control over the immense, growing, and dangerous drug industry. We provided Blackhawk helicopters, defoliation spray planes, military training units, and much more.

The US government had a special interest in bringing down the drug cartel, because Escobar's Medellin Cartel was responsible for about 80 percent of all cocaine coming into the United States. Despite Colombia's best efforts, the Medellin Cartel's power grew exponentially in the late 1980s.

Pablo Escobar began as a street criminal, stealing headstones from cemeteries, then stealing cars, and gradually rising to become one of the world's wealthiest men. *Forbes Magazine* ranked him the seventh-richest person on Earth, having a personal fortune of about $24 billion. His take over of the Medellin Cartel began in 1975 with the death of Fabio Restrepo—who was probably murdered by Escobar.

Pablo called his simple business model *Plato o Plomo:* Silver or Lead. He would attempt to bribe officials (silver). Anyone who didn't cooperate was murdered (lead). Police were frequent targets, along with politicians, elected officials, and competitors. Escobar bombed a municipal police headquarters and destroyed a commercial airline flight, killing 110 innocent people in an attempt to assassinate a presidential candidate who missed the flight. Escobar could order the murder of anyone, at any time, in any place. His tentacles reached worldwide, and he was personally responsible for killing up to 4,000 people.

When Escobar took over the cartel, smuggling consisted of carrying small quantities of cocaine in suitcases. This method was profitable, but small scale. Labs in the jungle could process and package a kilo of coke for about $1,500—and on American streets that same kilogram would bring in $50,000. Former dope smuggler Carlos Lehder devised a plan to use small aircraft to smuggle larger quantities. When he joined Escobar, the profits became nearly unimaginably huge. They invested in improving the technology, and staying one step ahead of the police and the US military. And the business grew more violent.

In 1989, Operation Centre Spike electronically followed Escobar around Colombia, using surveillance equipment accurate enough to pinpoint a telephone transmission on the ground to within 200 meters. Beechcraft airplanes, designated as C-12s in the military, were equipped with specialized electronic equipment and personnel who used computers and sophisticated technology to eavesdrop on Escobar's telephone conversations—not only locating the man, but building evidence of his activity. These flights continued over Colombia until 1991, when the planes and people were pulled out and moved to Somalia.

Operation Centre Spike ceased in 1991 because Pablo Escobar was in jail—a prison of his own making. Fearing he'd be extradited to the United States, Pablo and his lawyer made a deal with the Colombians. He would turn himself in and accept a five-year sentence in prison—if he could build the prison himself. He called it *La Catedral*, a walled fortress with a soccer field, bar, pool, and other luxuries. Escobar was the only prisoner, and he continued to operate his expanding drug and terror business from there. Stories also abound that he murdered disloyal employees inside the prison.

Because Escobar was "behind bars," the Centre Spike planes and personnel moved on to other assignments. Initially, I flew to Mombassa, Kenya, to pair up with another pilot. After a week or so in Kenya, it was time to go to work in Somalia.

We landed the plane in Mogadishu and found ourselves in primitive territory. Base camp sat smack in the middle of the airfield, surrounded by fences and guards all around the base. We lived in tents and ate food from a military chow hall. We wore helmets and flak jackets, because things were heating up in the town just outside the perimeter.

Flying in Somalia was different. We had no air traffic controllers other than the military tower at Mogadishu. After departure, we tuned up a common frequency and all separation of traffic was done pilot to pilot. This wasn't a problem, because the weather was always nice, and few planes were flying. Most of my flights entailed carrying command and staff personnel to outpost units stuck out in the middle of nowhere, where soldiers lived in tents. The conditions were awful for these guys. It was hot and humid, and the flies were thick. I felt lucky to be living in Mogadishu in my air-conditioned tent with hot meals served at dinner time.

I had a great time flying to these outposts, because we got to land our C-12 on dirt roads and unimproved fields. One site was especially interesting because the unit had scraped out a small 2,500-foot dirt runway. We landed and loaded eight soldiers with all their gear. We didn't need performance charts to tell us the weather was too hot and the runway too short. Major command gave us permission to operate the aircraft however we felt—accomplishing the mission was our standard. In combat, all rules are thrown out, and you do what you have to.

Although the runway was short, we had miles and miles of barren field beyond the end of it. If we didn't get off the ground, we'd roll to a stop in the field, probably wiping out the props with sagebrush. We taxied to the end of the runway, held the brakes, ran the power to max, and began our takeoff roll. All the tired, smelly soldiers in the back let out a cheer because they were going back to Mogadishu for much deserved rest and relaxation. Little did they know, the two pilots up front were sweating profusely.

Our takeoff roll progressed slowly, and we were a long way from our liftoff speed of 98 knots. As the end of the dirt runway approached, I looked at my airspeed. We were way too fast to stop and too slow to lift off. When I reached the end of the runway, I desperately heaved back on the yoke as hard as I could. Far below liftoff speed, the plane actually jumped off the ground. As soon as I got the main gear off the ground, I lowered the nose to keep the plane in ground effect and retracted the gear to reduce drag. We floated at least a thousand feet in ground effect, with the stall horn blaring, until we built enough airspeed to climb out. The whole time, we were on the ragged edge of a stall, dangerously close to nosing over into the ground. When the speed built to a minimum needed to climb, I retracted the flaps and we flew on to Mogadishu. The guys in back were clueless, thinking of soft cots and warm food.

When things blew up in Somalia, our fixed wing support in Mogadishu ended. We flew the plane back to Fort Bragg, by way of Saudi Arabia, Crete, Naples, Germany, Scotland, Iceland, Greenland, and Goose Bay, and then on to Maine. That's when I got the call to start my new job near Washington, D.C., and from there to Operation Centre Spike in Colombia.

Pablo had escaped from prison.

In mid-July 1992, Escobar left his prison and vanished into the Colombian countryside. Massive manhunts were launched, including government police; a group called the Search Bloc, which was a US-trained Colombian task force; *Los Pepes*, a loose organization of Escobar's enemies who were financed by the Cali Cartel; and me. Operation Centre Spike began anew, using airplanes against the very cartel that developed aircraft smuggling into North America.

Based in Bogotá, our main area of operation was over the Medellin area. We never had a chance to explore this beautiful city, because one never knew who was on Pablo's payroll, officially or through bribes. Bogotá was risky enough. Trying to operate in the city controlled by the Medellin Cartel would have been suicide. Mostly, I saw the country from the air. We'd take off from the main airport under the guise of a corporate aircraft. The specialists would plug in their laptops, and we climbed high to fly grids over selected areas of the country for hours at a time. By triangulation, our equipment and the surveillance groups on the ground could locate a phone within inches. Our techs in the cabin, knowing Spanish, would find our quarry through a telephone conversation, then contact ground surveillance teams to give them the target's exact location. Sounds easy, but it was not.

With all of this technology, finding Pablo was still a huge challenge, because informants tipped him off every time the US and the Colombian military tried to move in. We would believe we had him cold, then the ground troops would find he'd moved.

The aircraft was the same model I flew in Somalia, with subtle, but significant differences. The cabin had special equipment installed, and the wings were six inches longer to conceal the main eavesdropping antennas. Other antennas located in the belly could be retracted and extended in flight, so while on the ground the planes looked like normal twin turbo-prop King Air business aircraft.

Most of the missions seemed fruitless, and nearly always frustrating. We would take off from the Bogotá airport, climb to 25,000 feet, and enter a holding pattern where we circled for up to six hours. From a pilot's point of view, this was boring work—most of the time.

One particular night was an exception, with terrifying life or death consequences.

Thunderstorms blanketed all of Colombia as we departed for our mission late at night, flying the usual profile. When the time came to head back to the airfield, we contacted Bogotá approach and were instructed to enter a holding pattern at 25,000 feet, with 13 other airplanes ahead of us for landing. We held for over 45 minutes while the stacked aircraft landed. As one airplane landed, approach control would clear all airplanes in the holding patterns to descend 1,000 feet. When a plane arrived at 12,000 feet, it was next in line for landing. To make things worse this night, one huge storm sat directly over the VOR, right where we were flying.

Not only were we pounded by turbulence—everything that wasn't tied down flew around the cabin—the combination of maneuvering around these deadly thunder cells and avoiding other airplanes created a tense pilot workload. The turbulence and heavy rain made using the autopilot impossible, so all flying was done by hand, an exhausting ordeal. Forty-five minutes seemed an eternity. I was sure we'd be struck by lightning or encounter extreme turbulence, maybe even hail. Any one of these things can destroy a plane. All three at once presented a catastrophic situation for us, but we were soon to meet an even more critical situation. We were low on fuel. The bad weather and delays in the holding pattern left us with little extra.

Slowly, we were moved down in the holding pattern. When we finally reached 12,000 feet, it was our turn for an approach to the runway. As we anxiously waited for the controller to give us the go-ahead, we got the bad news: the airfield had closed because heavy rains caused a power outage. All lights on the airfield were out, and the runway was covered by over six inches of water. We, of course, couldn't see any of this because we were being hammered by turbulence, blinded by lightning, and unable to view anything beyond our wings because of dense clouds. Landing there was impossible. We had to make a quick decision or we'd be dead.

The decision was made for us. We had no option. The only place in Colombia with acceptable weather was Cali. There was no other place to land. None!

Almost exactly a year before this horrible night, I faced a similar emergency in South Korea while flying an RC-12 spy plane. I was a unit trainer, orienting new pilots in the mission and flying intelligence-gathering flights along the border with North Korea. We conducted five-hour missions 24 hours a day, seven days a week, rotating several planes to keep a least one aircraft in the air at all times.

On one of these missions, dense fog socked in the entire Korean peninsula and around two o'clock in the morning, I began eavesdropping on the weather reports back at our base, Camp Humphreys. Things sounded grim. Visibility had decreased all evening, with fog in the forecast—exacerbated, as always, by the burning of *ondal*, a block of charcoal the Korean people used to heat their homes in winter. I knew it was time to break track and head home before visibility went below landing minimums.

The flight back was about 30 minutes, and in that time the flight visibility went from two miles to less than one-quarter mile. I frantically began calling other airfields in Korea within my fuel range. All of them reported visibility below one-quarter mile. Korea was socked in. I had no option but to execute an approach and hope like hell we could see something. Just like the night over Cali, fuel became a critical problem. If we didn't get in the first time around, we probably wouldn't have enough fuel for a second approach.

We tuned up the ILS frequency and began the approach into Camp Humphreys. Lucky for us, the winds were calm and the air smooth. As we continued the approach, the controller told us visibility was so low the tower couldn't even see the runway. Neither could I. I asked them to turn the airfield lights full bright and wish us luck.

We continued down to 200 feet, landing minimum, and still couldn't see the glow of runway lights. At 100 feet I spotted a faint glow, so I began slowing my airspeed and bringing the nose of the plane up. At

50 feet I was at landing speed and certainly in the landing attitude. All I could do was hold the nose of the plane slightly up in a landing position. If we hit the ground, at least we would hit with our tires first.

At 25 feet, my copilot said he could barely make out what looked like runway side marking lights. I pulled back on the yoke, pulled off the power, and, amazingly, felt a slight shudder and heard the light squeak of the main tires contacting solid ground. What a feeling. I put the nosewheel on the ground and pulled the power levers into reverse thrust. We stopped somewhere on the runway, and I set the brake. I turned and looked at my partner, and we both broke out laughing, guffawing to the point of tears. It was the best landing I've ever made in an airplane. I made the landing without ever seeing anything besides the faint glow of runway lights. It was so foggy we couldn't even taxi to the hangar, and ground personnel took 30 minutes to find us so they could pull us in with a tug.

In Colombia, we had a slightly better option. Leaving the holding pattern at Bogotá without hesitation, we climbed back to altitude and headed for our alternate airport in the city with the second largest drug cartel in the country.

The flight wasn't long, but it quickly ate our remaining fuel. When we were within radio range of Cali control tower, we called and requested landing clearance, fully expecting it would be given. Were we ever wrong!

Because we hadn't filed a flight plan to their airport, Cali tower denied our request to land. My second request included the word *emergency*, and the fact that we had no choice. The controller refused to budge. I declared an emergency and he still denied me.

Under no circumstance was the city of Cali, Colombia, going to let an unknown airplane with English-speaking pilots land without prior permission. Once again I requested landing clearance and explained the entire situation to him. He didn't care; we weren't going to land at his airfield.

I told him we *were* going to land, and in one of two ways: first as a twin-engine turboprop, or second as a glider. We were running out of fuel. He didn't think I was funny and stopped answering my radio calls. Silence from his end.

I wondered whether the Cali drug cartel had bribed airport officials, and probably the police, to keep foreigners out. The cartel certainly was powerful enough to make the decision on who came and went on their turf. I was certain all the cartels knew about our operations, or at least that we were in country trying to slow the drug flow to America, their biggest market. If we did land, with or without permission, what would be our fate?

If the Cali Cartel was in control, we'd probably be killed, especially if they saw the equipment inside the plane. We could die by running out of fuel, or after landing we would die at the hand of some cartel assassins.

Then it occurred to me: The Cali Cartel actually was an enemy, a competitor, of Pablo Escobar and the Medellin Cartel. If the Cali people knew we were searching for Pablo, maybe they'd go easy on us. But our best chance was to maintain our cover and deny any involvement in the anti-drug trafficking programs. We were a corporate aircraft. I pushed those thoughts out of my head, mentally reviewed my cover story and turned our plane toward the runway. In these situations, focus is critical.

Cali wasn't a busy airport in the middle of the night, but diversions from Bogotá brought other planes into the area. With no other option at hand, we simply lined up on the landing runway, dropped our landing gear, and began an approach to the runway. The landing was smooth and a rush of relief ran through our minds as we rolled to a stop on a taxiway. Although we averted disaster in the skies, we now faced a new set of circumstances.

I shut down the engines, knowing the guys in the back were scrambling to secure their gear and probably deleting files. Several trucks loaded with military personnel carrying rifles were already rolling down the taxiway, moving quickly toward us. They circled our airplane, and the troops dismounted with guns in hand. Spotlights from several jeeps blinded us, but we knew they wanted us out of the plane. I unstrapped my shoulder harness and headed back to open the cabin door, fearful of what waited outside. I figured it was better to meet these guys outside the plane, away from our sensitive equipment, and in an area where we had room to move if attacked.

Four of us were aboard the plane that night—two pilots and two intelligence officers. Only the two Intel officers spoke Spanish, so we depended on them to talk our way out of this mess.

However, the Colombians weren't interested in talking. As soon as I opened the door stair, several men reached in and grabbed me, one clutching my long, shaggy hair. Other soldiers pulled my copilot and the Intel guys out. They tossed me onto the ground, smashing my face into the hard surface. Gravel and small pebbles dug into my cheek. With my face pressed into the pavement, I felt a boot on the back of my neck, pushing hard, grinding my face on the ground. A rifle barrel pressed against the back of my head. It hurt like hell. We were in deep trouble.

For the first time I was second-guessing my decision as captain to divert to Cali. Maybe we should've taken our chances crash-landing in the jungles of Colombia in the middle of the night. Thoughts of torture flashed through my mind. These Colombians would want to know what we were doing in their country, although I'm sure they had some idea. The entire time I lay on the ground with the rifle digging into my head, I went over my cover story so that, when questioned, I wouldn't stumble or sound like I was making up a lie. We had an elaborate cover, and I needed to relate my story without error. One mistake and the Colombians would dig deeper to find the truth.

A young soldier pulled my hands behind me, arching my back and straining my neck. I felt blood running down my forehead. With my head pulled backward, I thought of photos I'd seen of people being assassinated in jungles from Southeast Asia to South America. I doubted these guys were any part of the organization, but I thought about FARC, a revolutionary group from the Bogotá area that started in the 1960s. They were involved in the cocaine business, using the trade to raise funds for their political activities. They were best known in the States for kidnapping and killing. With this thought in mind, I imagined the Colombian with the rifle at my skull reaching down, pulling my head further back by my hair, and firing a round into the back of my skull.

Instead of shooting me, my captor relaxed his grip a bit. He didn't let go, but I felt the slight loosening of tension. I wondered how my crew were holding up, whether they also were bleeding.

The four of us from the plane lay on the taxiway for at least twenty minutes before another jeep arrived. The engine stopped. I turned my head slowly, wanting to see what horror was coming next. All I saw were shiny boots and a pressed uniform with a holster and pistol. *He must be the boss,* I thought. This obviously was a high ranking officer. He dismounted and the troops around him seemed intimidated by his presence. He strolled around the airplane and then began mumbling to himself. I had no idea what he was saying, but the word *gringo* came up several times. Not too difficult to understand who he meant.

Suddenly, he broke into perfect English and said, "What are you doing in my country?"

I felt like I was responsible for this situation as aircraft commander, so I opened my mouth to reply. Before I could say a word, one of the intelligence officers began speaking in Spanish. I felt my captor tense, and I tried to pull my head away from the rifle muzzle.

I turned my head, risking getting poked harder, so I could see the Intel guy who had spoken. A middle-aged soldier roughly pulled him to his feet. The CIA guy's hands were free so he kept his balance— and somehow managed a calm demeanor. The soldiers stayed close to him, keeping weapons aimed at his torso. The spy and the Colombian commander walked about 50 feet away. Both spoke Spanish, and their conversation dragged on for twenty minutes. I suspected the man guarding me was watching them too, because he relaxed again. I turned slightly so I could see the two men from the corner of my eye.

I don't know how long this went on, but at least three times during the conversation the Colombian commander lit and smoked a cigarette. At times, the conversation became heated. Other times, it seemed relaxed and almost amiable. Either way, the time lag gave me too much leeway to consider what might come out of this ordeal. I had trouble seeing anything positive whatsoever.

This could become a huge international incident if our cover was blown. The Colombians could try to make an example of us. Even worse, they might hand us over to the very cartel we wanted to eradicate. In fact, I thought, they might well be on the cartel's payroll. If so, I couldn't imagine a positive outcome. I knew what Pablo Escobar did to informants and cheats. For the first time in my military career, I thought I'd have to use the extensive training I received in handling torture.

A trickle of blood pooled at the corner of my eye. My thoughts of doom ended when I saw the conversation between the two come to an end. To my surprise, they shook hands, which had to be a good sign. As the spy and the commander walked toward us, the Colombian officer barked orders and the soldiers jumped to attention. They withdrew the rifles and removed their boots from our necks so fast I bumped my chin on the pavement. Without another word, the soldiers loaded back into the trucks and drove away from our plane.

I don't know how long I'd been lying on my stomach, I felt stiff and sore along my spine and legs. I crouched on my hands and knees, rising slowly. As I looked up, the sight was amazing. A fuel truck moved down the taxiway toward our airplane. Could this be real?

The Colombian officer was still around, and I didn't want to say anything that would change his mind about letting us go. We filled up with fuel, climbed aboard the airplane, and started the engines. The sun was coming up, and the weather had improved significantly. Cali tower issued us clearance to depart, and we took off without even filing a flight plan. As we flew back to Bogotá, the four of us sat in near silence. The whole ordeal seemed surreal.

It wasn't until we landed in Bogotá that I finally found out how the Intel officer talked our way out of trouble with the Colombian commander.

"No matter what I told the Colombian officer, nothing mattered. He would argue and threaten to hold us," explained the Intel officer. "That is, until I mentioned we carried large sums of money to pay for fuel."

The money was actually "contingency" money to be used to bribe our way out of trouble. It worked.

"That's probably the most expensive fuel ever purchased," he concluded.

The next day we were debriefed at the embassy and went off on another mission, still chasing Pablo.

Everyone knew Pablo Escobar liked young girls and married his wife when she was only 15. Though Pablo was unfaithful, he tried to protect her and his children. As the situation became more risky for him, Pablo tried to get his family out of Colombia. He became agitated with the government for interfering with those efforts.

Escobar knew we were tailing him from the air, but he still used his mobile phone on December 2, 1993, to telephone a radio program and complain about treatment of his wife and children. Operation Centre

Spike listened from the C-12 aircraft above, and we nailed him with this call by using triangulation.

The Colombian police moved in. Pablo Escobar made a run for it, leading to a shootout on the rooftop of the building where he'd been hiding out. Escobar was killed, shot in the leg, the torso, and the head, the last bullet entering through his ear.

To this day, we don't know who put the bullet in Pablo's head. Some say he took his own life. Others say US Delta Force killed him, and still others say the Colombian police were responsible.

We'll never know. But I do know that after the Medellin Cartel fell, the Cali Cartel moved in—and these folks were more sophisticated and dangerous than Pablo Escobar.

Pablo Escobar is gone and the fantasy park he created and the wildlife from Africa that roamed the estate grounds are history—except for the hippos. They are thriving. And, no thanks to the Cali soldiers, so am I.

Nick inside a captured Iraqi bunker, 1991.

1991 in Desert Storm, after a successful air attack.

In the cockpit of a Mohawk in South Korea,
Fort Humphreys 1992.

Nick (on left) with the RC-12, twin engine plane similar to that used in Colombia during Centre Spike.

Nick with Bedouins in Iraqi desert, 1991

The Ice

A Personal Account of Flying in Antarctica
by Captain Paul Derocher, US Navy

"Mayday!" That's one word no pilot wants to hear—especially when the voice is someone you know. The emergency distress call came from my best friend, and he had just crashed on a remote glacier in Antarctica. Luckily, I was only minutes away, flying supplies to the South Pole. I immediately turned my plane around and was soon flying low over the crash site.

All of the crewmembers were standing on the glacial ice near the heavily damaged aircraft, waving their arms as I flew by. My buddy, the copilot on the plane, was transmitting to me on his hand-held emergency radio, telling me there were no injuries, but I could see they had been very, very lucky. The C-130 Hercules ski plane, a large military transport aircraft about the size of a Boeing 737, was precariously straddling a crevasse so wide and deep that the entire aircraft could have been swallowed whole. The nose and tail of the aircraft were on opposite sides of the crevasse, while the mid-section of the fuselage spanned an icy abyss. Two of the aircraft's four engines were ripped off the wings, one careening into the left side of the fuselage and the other lying off the right wingtip. Red hydraulic fluid from the wounded aircraft was bleeding onto the glacial whiteness.

Only a few minutes before, the C-130 ski plane had landed on the Starshot Glacier to unload a group of scientists and their equipment. Looking down from my cockpit, I could see the track of the stricken aircraft's skis on the glacial ice: the aircraft's straight track in the snow where they landed, unloaded the scientists, and then tried to take off. At the end of the mile-long straight track, the aircraft's trail became a broad sweeping curve across the snow to the crash site. The ski track showed that the aircraft had taxied across not just one, but a series of ever-wider crevasses before it finally reached one too large to cross. Almost unbelievably, these crevasses were all but invisible to the pilots as they slid across them.

 I radioed back to the crew on the ice not to move around. They were in the middle of a spider web of crevasses, and if they were to walk in any direction they might fall into one of the massive, deep fissures in the ice—almost certainly to their deaths. Their situation was perilous. The crew had escaped from the aircraft but without their survival gear; some of them weren't even wearing a flight jacket in the -20 degree temperature, but the stricken aircraft's stability was so uncertain that no one could risk going back inside to get their tents and emergency rations.

 The seven-man crew had to be rescued quickly before they perished from exposure in the bitter Antarctic cold, but it would have been suicidal to land my own C-130 Hercules ski plane anywhere near them. The science team was about two miles away, but they, too, would be unable to go to the crash site because of the deadly crevasses between them and the aircraft crew.

 I radioed our primary base at McMurdo to launch a rescue operation. Within minutes, two of my squadron's Huey helicopters began their 300-mile flight. I orbited the crash site for nearly three hours before the two Huey helicopters came into view, flying up the Starshot Glacier at low altitude. They flew straight to the stricken aircraft and landed between two crevasses, next to the seven men and the badly damaged Hercules. As the rotor blades whirled over their heads, the C-130 crewmembers crouched down and ran through the cloud of blowing snow into the helicopters. Within minutes, both Hueys lifted off and flew down to the frozen Ross Ice Shelf and back to McMurdo.

 A few days later, I was back at the crash site as the leader of the accident investigation team. We flew to the Starshot Glacier in one of my squadron's helicopters. Prior to landing, we flew along the track of the aircraft across the crevasses, then hovered over the crevasse the Hercules ski plane was straddling. From the air, I peered straight down into the canyon of ice. It was so deep it was frightening to think about going anywhere near it, but there was no choice. That's where the airplane was.

Although I was the only member of the accident investigation team who had any experience climbing in crevasses, I was by no means an expert. I had only learned how to ice climb in Antarctica the year before, so my experience was limited. As a member of my squadron's para rescue team, I climbed into and out of the steep crevasses near our base in McMurdo, practicing rescues of people who fell in.

We got out of the helicopters, put on our ice-climbing harnesses and walked slowly toward the crevasse. I repeatedly pushed my ice axe full length into the snowy surface, probing for more hidden crevasses. When we finally approached the lip of the giant fissure, I lay flat on the snow, crawled to the edge, and peered over.

This crevasse was much deeper than anything I had ever climbed before. Holding one end of my extra climbing rope, I tossed it into the opening to get a rough sense of the depth. The rope almost entirely played out. The crevasse was about 150 feet deep. The aircraft was only a hundred feet long and could have fallen completely into the crevasse, likely killing everyone aboard.

To the left and right of the opening, the top of the crevasse was entirely masked by a weak crust of snow only a few feet thick—called a "snow bridge"—that was completely level with the glacier's surface. It stretched for a thousand feet to either side. As I looked down into the crevasse under the snow bridge, I could see the gloomy subterranean fissure stretch away to the left and right into deep dark blue caverns. Very deep and with nearly vertical sides, the crevasse was eerily quiet and intensely cold.

After assessing the situation at the crash site, I flew back to McMurdo, where I met with the navy's senior officer, the commander of Operation Deep Freeze. We decided to salvage the aircraft. That meant we had to somehow pull the aircraft off the crevasse and then repair the extensive damage to the wings and engines. To do this, a large campsite had to be constructed near the crash site, where our mechanics, technicians, and engineers could live while they repaired the aircraft.

The Starshot Glacier was a hazardous place to work. The glacier was flowing from the mile-high ice sheet down through the valley to the

Ross Ice Shelf. New crevasses were being formed—somewhere. Using our helicopters for aerial surveillance and skilled mountain climbers from McMurdo, we located a safe place on the Starshot Glacier for our campsite and a landing area suitable for C-130 ski planes. We then started flying in the necessary people and supplies. To pull the aircraft clear of the crevasse, we also flew in a D8 Caterpillar tractor, which had oversized tractor treads to keep it from sinking into the snow.

The first step was to pull the aircraft off the crevasse. The civil engineers on the Operation Deep Freeze staff—the Seabees, and the National Science Foundation's civilian contract engineers—devised a plan to pull the aircraft clear. The theory was that when the Caterpillar tractor's 18-ton weight was attached to the front of the aircraft and started pulling it, the weight of the tractor at the nose would be heavy enough to keep the tail of the aircraft from sliding backwards into the crevasse once the tail of the aircraft was no longer supported on the opposite side. But nobody had tried anything even remotely like this before.

We positioned the D8 tractor in front of the aircraft's nose, and attached a tow bar from the tractor to the nose landing gear strut of the aircraft. On the other side of the crevasse, a very large airbag was positioned under the aft section of the fuselage, which was resting on the ice. The tractor's engine was already running when we started slowly inflating the airbag under the tail of the aircraft. Then, something unexpected happened—the aircraft started sliding forward towards the tractor on its own.

The aft section of the fuselage of the C-130 slopes upward at a steep angle. Because of this geometry, when the airbag under the aft section of the fuselage started to inflate, part of the force of the air filling the bag not only pushed against the tail above it but also pushed the aircraft forward towards the D8 tractor. It was simple trigonometry but no one was breaking out his calculator to compute the exact force. The tractor driver sensed the motion and instinctively started towing the aircraft clear without awaiting any kind of orders. Within no more than two minutes, the aircraft was pulled well away of the crevasse. What was expected to be the hardest part of the salvage turned out to be the easiest.

For the next several days, the mechanics in my squadron removed what was left of the two damaged engines and replaced them with two new ones. Severed fuel, oil, hydraulic, and pneumatic hoses and ducting were repaired. Within three weeks of the crash, the aircraft was flown off the glacier and back to McMurdo.

Glaciers and Crevasses

Antarctica is a treacherous place to be. In my three years in Antarctica, 1983-1986, two of my squadron's 12 aircraft crashed on glaciers and mountains, and a ship was crushed by ice floes and sank. We were fortunate—no one died when I was there—but we'd been lucky. Within a few years after I left, several friends were killed in aircraft crashes in Antarctica. Two others died when falling into a crevasse while walking near the primary US base in McMurdo.

What is it that makes flying in Antarctica so different and so dangerous? Antarctica has the world's coldest temperatures and strongest winds, but the worst dangers are crevasse-riddled glaciers and fatally disorienting visual illusions that render mountains nearly invisible even in clear air. Only a few years before I arrived in Antarctica, 257 people were killed in a whiteout when their aircraft crashed near our base at McMurdo.

To survive flying in Antarctica, you need to know the nature of the land itself—or rather the ice. In fact, Antarctica has been called The Ice since the days of the early explorers, with its overwhelming vastness of ice in its many forms: ice sheets, glaciers, snow, ice shelves, and icebergs. Every winter, the surrounding ocean freezes for hundreds of miles beyond the coastline, more than doubling the surface area of the frozen white continent. Ninety-eight percent of Antarctica is covered under a vast glacier that averages well over a mile thick. The South Pole, for example, is at an elevation of over 9,300 feet, and it is solid ice all the way to sea level. This is such a vast amount of ice that if it all melted, sea levels throughout the world would rise about 150 to 200 feet.

Antarctica is about the same size as the United States and is roughly centered around the South Pole. It is often compared to Alaska but it is far colder, with the world's coldest temperature of 128 degrees Fahrenheit below zero. Unlike Alaska, there are not only no towns, but not even any plants or land animals. The only life that exists is birds and other animals that live in the sea, such as seals or penguins.

Under the force of gravity, this continent of ice flows inexorably downward and outward to the surrounding ocean, sometimes at the rate of several feet a day. Long, floating tongues of ice are formed in areas where the mile-high Antarctic Ice Sheet flows through mountain valleys into the sea. At other places, these rivers of ice ooze into bays that somewhat restrict the seaward flow, so that streams of ice merge to form vast, floating ice shelves. The ski plane runway at the American base at McMurdo, for example, is located on the Ross Ice Shelf, a floating glacier about the size of France. The ice shelf's seaward edge is a sheer cliff, often a hundred feet high, which extends for hundreds of miles. When the seaward edge of the ice shelf fractures and breaks off, vast flat-topped icebergs are formed that are hundreds of square miles in size.

When I flew in Antarctica in the early 1980s, much of the interior was incompletely or inaccurately charted. Even mountains close to the primary American base at McMurdo were depicted on various charts with elevations that differed by nearly a thousand feet. Throughout the continent, most of the mountains and other geographic features remained unnamed. We in the squadron often flew to and landed at sites where no human being had ever set foot.

And I think that bears repeating—we were the first human beings to ever be there in the history of mankind. For me the appeal to fly there was irresistible despite the risk.

So how and where do crevasses form and why can they be so difficult to detect? In mountain valleys, glaciers are slowed as they flow along the mountains on either side. Typically, a series of more or less parallel crevasses will form at about a 45-degree angle to the sides of the valley. For a pilot, landing close to the mountainside on a glacier is often unsuitable, but that doesn't mean the middle of the glacier is any safer. Crevasses can also form midstream as the glacier flows over uneven, down-sloping terrain on the valley floor. The surface of the glacier may be forced to bend downwards and if the ice bends too much, it starts to crack open. Crevasses can also form where the glacier makes a turn through the valley or if it merges with another glacier.

Crevasses start out as hairline fractures and the sounds of these "ice quakes" can be heard, especially in the silence of the night. Often as I lay on the surface of the ice in a sleeping bag on remote glaciers, I heard the growling as the glacier made its way down the valley. As the glacier continues to flow, the ice on the surface separates even further on either side of the fissure, becoming deeper and longer along the surface. Like the crevasses that form along the sides of the mountains that contain the glacier, crevasses in the middle of the glacier form more or less parallel to each other. However, the stress of flowing along an erratically uneven valley floor can cause two or more patterns of crevasses, sometimes at right angles to each other, creating square-sided towers of ice as high as a ten-story building.

If this were all there was to it, crevasses would only be an obstacle to movement on the surface, not a potential killer. You would see the crevasse and simply go around it. The problem is snow bridges, which can completely mask the crevasse below but be too weak to support your weight. The term snow bridge can be misleading. When we think of bridges across a river, we visualize a narrow connection between two sides of a river but a snow bridge is not a narrow pathway at all. The entire width and length of the crevasse is bridged. On the surface of the glacier, the snow bridge appears as smooth as the snow on either side of it, with no indication that a crevasse is below.

Here's how they form. When the surface of the glacier fractures, the fissure may initially be only a few inches wide. As the wind blows snow along the surface of the glacier, the snowflakes will stick to both sides of the crevasse, adhering like bits of Velcro. The snow continues to accrete on both sides until they meet, bridging the crack. As the crevasse widens further, the winds continue to blow more snow, the snowflakes continue to adhere to both sides of the crack and to each other, and the snow bridge widens. Snow does not fill the crevasse.

The snow bridge itself is far weaker than glacial ice on either side of the crevasse. The strength of each snow bridge varies along the length of the bridge's surface. Sometimes, the first person walking across the crevasse will make it to the other side, but then the snow bridge can collapse when the next person following behind steps onto it.

Underneath the snow bridge, the crevasse typically presents two nearly vertical cliffs of ice, which taper as they get deeper. If you fall into a crevasse, you might be wedged so tightly that you can barely breathe. That's what happened to the two hikers from McMurdo who fell into a narrow, snow-bridged crevasse.

Operation Deep Freeze

When I was in Antarctica, all of the flying within the continent of Antarctica was flown by the US Navy. We were the air arm of the US Navy, aptly named "Operation Deep Freeze," which coordinated all the military air and sea transport operations.

My squadron was composed of seven ski-equipped C-130 Hercules transport aircraft and seven UH-1 Huey helicopters. I flew both aircraft. My squadron was one of the largest deploying squadrons in the navy. About 400 men and women flew and maintained the aircraft or provided support services. I was their commanding officer and, to be frank, my overriding concern was to get everyone home alive.

The primary base for the US Antarctic Program was McMurdo Station, which we called "Mac Town." It was located on Ross Island about 40 miles from the mainland of Antarctica itself. Ross Island was dominated by the 13,000-foot (more or less) high Mount Erebus, an active volcano that was continually emitting smoke and lava "bombs." During the Antarctic summer season, Mac Town was home to nearly 1,000 scientists, navy flight crews, mechanics, technicians, and engineers. Mac Town's no-frills barracks, galleys, oil tanks, and warehouses resembled mining towns in Alaska and the Canadian Arctic. Building insulation was so thick that freezer doors were used for doors to the outside, but it was the outside that was the freezer—not the inside. The "refrigerated" warehouse for fresh food flown in from New Zealand actually used a heater to keep the temperature inside the refrigerator from freezing and spoiling the fresh vegetables.

Within the first few weeks of our arrival in Antarctica, all flight crewmembers went to polar survival training. The course was taught by the NSF's contracted mountaineers, including my friend Rob Hall. We spent a day on the Ross Ice Shelf digging ice cave shelters, building igloos, pitching tents, and learning how to cook the survival rations. We wore the same gear as when we flew.

Our first layer of clothing was long john underwear, followed by a thick insulated vest and pants, and finally a flight suit that had to be four sizes larger than normal. We looked like the Pillsbury Doughboy. Our olive drab Korean War-vintage parkas kept us warm but were very heavy, especially compared to modern winter clothing. Our long, thick gloves were covered in fur and were called bear paws. We Herc pilots wore standard issue flight boots when flying, but when flying the Huey helicopter, we wore the soft, felt-lined Eskimo style footwear, called the mukluk, to give us added feel on the rudder pedals.

Operation Deep Freeze worked for the National Science Foundation (NSF), which directs the United States Antarctic Program. While global warming research and ozone hole studies get much of the media attention regarding Antarctica, the NSF funds a broad spectrum of other scientific research, including upper atmospheric physics, meteorite studies, seismology, glaciology, geology, and marine biology. Millions of dollars in grants are awarded to the scores of world-class scientists in NSF's Division of Polar Programs.

I've had a life-long passion for science and received a NSF grant myself years before to study astrophysics, so I especially enjoyed the opportunity to learn about the many science projects being conducted in Antarctica directly from the principal scientists heading the studies. At the US Naval Academy, with its heavy emphasis on mathematics, science, and engineering, I majored in oceanography, a major area of study in Antarctica. Every Sunday evening, scientists would brief us on their projects at the NSF's headquarters in Mac Town. The presentations were open to everyone and were nearly always standing room only.

Activities at McMurdo were limited, but a few of us didn't let the extreme cold prevent us from going outside. I've been a lifelong runner, but the four-mile round trip up and down the steep icy road from McMurdo to Scott Base was a challenge even when I was young. The first Antarctic marathon started while I was there, but I was only able to slog through the snow and ice for about a third of it before I was so numb I couldn't feel my feet stepping on the ice.

The New Zealanders at Scott Base had jury-rigged a rope tow on a nearby slope where you could ski downhill, if only for a short distance. The Kiwis, as the New Zealanders call themselves, had even built a little chalet at the bottom of the slope, a great meeting place that we jokingly compared to the bar scene in the movie Star Wars. For me, it was especially easy to mix with Kiwis. Only a few years before, I had served with the Royal New Zealand Air Force, flying the P-3 Orion in Auckland, New Zealand. I enjoyed the Kiwis' warm beer and dry humor.

It was at the ski chalet where I first met and became good friends with Rob Hall. Tall, lanky, and constantly laughing, Rob was a world-class mountaineer from Christchurch, New Zealand, and over the years I often hung out with him and his friends in Antarctica and back in Christchurch. Despite his relaxed demeanor, Rob was an exceptionally skilled mountain climber who often accompanied science teams into the Antarctic wilderness. It was Rob who introduced me to ice climbing.

I don't want to leave you with the impression that being in Antarctica was nothing but fun and adventure. One of my crewmembers said the whole operation in Antarctica reminded him of the TV show M.A.S.H. (I hope he didn't think I was Major Burns). Our uniforms were heavy, bulky, Korean War-vintage parkas, and our people in the field slept in tents. We were limited to two showers a week and the showers could only be two minutes long because it took so much fuel to melt the ice for bathing and drinking. The 24-hour sunlight each day made it difficult for me to sleep. When the sun's up, so am I.

The hardest part of living in Antarctica was the feeling of isolation, especially for people at the remote camps. The pilots who flew the C-130s and Hueys were able to see much of Antarctica on a daily basis, but most people were virtually confined to their station. They didn't even get to see any penguins, because the nesting grounds, called rookeries, were many miles away from McMurdo, accessible only by helicopter.

The vast distance from the United States meant that mail took weeks to get back and forth, and in the days before the Internet everyone monitored inbound flights from Christchurch to see if there was mail on board. Adding to the sense of separation, our five-month deployment schedule from October to February meant we missed both Thanksgiving and Christmas every year, as well as other personal and family occasions that were even more important. These were the same dispiriting conditions that all military families face with overseas deployments in war and peace, and at times it was just short of depressing. Luckily, there was little time off.

The Ice Runway and Skiway

In Antarctica, there are no runways, per se. The primary landing area during the first several months of our annual deployment was made entirely of frozen sea ice, approximately ten feet thick and strong enough to bear the weight of very heavy aircraft. We called it the "ice runway," and it floated above several hundred feet of ocean.

Our other "runway" was at Williams Field on the massive Ross Ice Shelf, located a few miles from both McMurdo Station and the New Zealanders' Scott Base on Ross Island. The Williams Field "runway" was composed of snow and could only be used by aircraft with skis, so it was called a skiway. The hard compacted snow is scraped relatively smooth with earth-grading equipment and outlined every thousand feet or so with black plywood panels. A line of red 55-gallon oil drums formed a crude arrow, pointing to the threshold of the skiway, to help you find the skiway when the weather was bad.

The people in my squadron kept aircraft flying in the worst weather on the face of the planet, yet there were no hangars for our aircraft at either the ice runway or the Williams Field skiway. All maintenance, including such complex and exacting procedures as engine removal and replacement, was done out in the open in temperatures as low as -40 degrees Fahrenheit. Many tools could not be operated while wearing thick, warm gloves, and our mechanics were forced to use thinner gloves or even work barehanded to get the job done. They could work for only a few minutes before going back to the heated shelters to keep from getting frostbite. Especially during the first few weeks of the flying season, the weather was both windy and extremely cold. In those conditions, no maintenance could be done. It was a challenge simply to fuel the aircraft.

Because the ice runway was as strong as a concrete runway, very large aircraft could land on it using wheels only—not skis. In early September each year, the US Air Force transport C-141 Starlifters started flying to McMurdo's ice runway from Christchurch, New Zealand. By mid-January each year the ice runway at McMurdo started to weaken and would not be safe to use for much longer. As the temperatures rose, but still well below freezing, the surface of the frozen sea ice would start to melt under the sun's direct rays. Flight crews and maintenance personnel would have to slog through ankle-deep freezing water to prepare our aircraft for flight. When the civil engineers determined that the sea ice was getting too warm, the ice runway was shut down. From that time on, all flying was done from the skiway. Our maintenance and other trailers, which were mounted on large skis, were towed from the ice runway to the ski way on the Ross Ice Shelf. Usually by April, the ice runway broke up into very large ice floes and was blown away by the winds.

Every five years or so, Williams Field would also have to be abandoned, but for entirely different reasons. Because Williams Field was located on the Ross Ice Shelf, the skiway was continually moving seaward where the Ross Ice Shelf would fracture into icebergs. This calving of the Ross Ice Shelf meant that all the buildings at Williams Field, many of which were mounted on large skis, had to be moved several miles back from the edge of the ice shelf and a new skiway scraped on the surface.

Living facilities at Williams Field were Spartan even by Antarctic standards. Dormitory buildings were trailers with insulated walls a foot thick and few windows. The building for taking showers was actually sinking into the snow at a crazy angle because the hot water flowing out the drains was melting the ice shelf below it. Williams Field had the honor of being the site for the annual Penguin Bowl football game. It wasn't a serious game, of course, with a remarkable hodgepodge of military and civilian winter gear, including the oversized Mickey Mouse polar boots. People drinking beer or soda in cans had to keep them in rubberized "cozies" to keep them from freezing.

Landing on Ice

The C-130 Hercules is a near-legendary transport aircraft, with its different variations serving for over 50 years in the US military. About the size and weight of a commercial airliner, the C-130 has four powerful engines mounted on its high wing. Although you see large propellers on the wings, each one is driven by a large jet turbine engine, hence the name turboprop engine. Lockheed Aircraft Corporation designed the Herc for tactical military missions, getting in and out of short landing strips in combat. The Herc could take the punishment of landing hard and taking off on rough unprepared surfaces, whether it was jungles in Southeast Asia, Mideast deserts, or glaciers in Antarctica. It isn't a glamorous aircraft. Neither sleek nor fast, the fuselage seems fat and the nose somewhat bulbous, but it was so tough we called it a tank with wings. I loved flying it.

Our Hercules had been specially modified with two 22-foot long snow skis, coated with Teflon, and mounted to the main landing gear on each side of the fuselage. A large ski is also attached to the nose landing gear. The Hercules skiplane can take off using either the wheels or the snow skis. Likewise, we could land on either the snow skis or the landing gear, which gave us the ability to land almost anywhere across the Antarctic continent. Other countries used much smaller aircraft—such as DeHavilland Twin Otters—with fixed skis mounted to the landing gear, but those aircraft had limited range and cargo capacity. The Russians used an aircraft about the size of the DC3 with oversized tundra tires rather than skis.

Our C-130 flight crews usually consisted of two pilots, a flight engineer, a navigator, and a loadmaster (who was in charge of passengers and loading cargo), but on the exhausting 22-hour round trip between Christchurch, New Zealand, and McMurdo, Antarctica, every position had at least one extra crewmember.

Taking off from the ice runway was the same as taking off from any snow and ice-covered runway back in America. We used our wheels, not the skis, to taxi the aircraft, and it was surprisingly easy to stop the aircraft on the ice runway by using brakes alone at low speeds. Using reverse thrust on the four turboprop engines also helped in slowing down but tended to blow snow in front of you, greatly reducing visibility. Frankly, I never knew of anyone on a C-130 or even the very large US Air Force C-141's who had trouble either taxiing or stopping his or her aircraft on the ice runway.

There was one interesting characteristic about the ice runway I hadn't been aware of until I saw one of our aircraft land. The ice runway was strong and also flexible. When the Herc landed, the ice runway visibly sank lower into the ocean. Because the ice runway was floating, it subsided into the ocean because of the weight of the aircraft. The ice in front of the aircraft actually rose, like the bow wave created in front of a large ship sailing through the ocean. Until I saw this, I thought I was just making very soft landings.

About halfway through the summer season, the ice runway was melting and unsafe to use. It was then that our C-130 ski-planes proved their true, indispensable worth in Antarctica. From then on, those ski-planes would fly all the flights between New Zealand and Antarctica—

to the delight of my crews—as well as within Antarctica.

Next to where we parked our aircraft at the ice runway was something not usually mentioned—going to the restroom. At the ice runway, our "facility" consisted of a large hole bored through the ice to the ocean below. On top of the ice hole was a plywood structure with a cartoonish (but tasteful) picture of a penguin sitting on a toilet and a sign emblazoned "Penguin Potty." It was a practical solution to an everyday necessity but with one unexpected drawback. On one occasion, one of our mechanics burst out of the Penguin Potty running with his pants literally around his knees because a seal had decided to come up the ice hole while our man was doing his business.

Flying the C-130 on Skis

Landing on skis was the same as for a conventional runway, right up until the skis touched the snow. Our principal concern was not making deep furrows in the snow that could make it hazardous for the next aircraft to land. If you made an abrupt turn when turning off the skiway, your tracks would pile up the snow so high that the next aircraft's skis might get stuck in your ruts in the snow, causing the aircraft to lean over so far that the wingtip or a propeller would hit the ice.

The goal was to make all turns gently. When I landed on skis, I could always feel the aircraft shudder as the skis impacted ridges of snow, even on the relatively smooth Williams Field skiway.

When I sensed the aircraft slowing down, I smoothly added power to the four engines to keep the aircraft moving and then turned off the skiway by making a very gentle turn. I kept the aircraft sliding over the snow until reaching a parking area clear of the skiway. Unlike most other aircraft, the Hercules had many windows in the cockpit, several of which were by my feet so I had a clear view of the terrain I was sliding across. You needed to get in the habit to keep checking how level the surface was where you were taxiing. Even at the well-maintained Williams Field skiway, a C-130 Hercules had taxied across a bumpy area years before, causing the wings to rock back and forth, hitting the hard snowy surface. The wing ruptured, opening up a fuel tank, and the aircraft was totally destroyed in the ensuing fire.

One thing you absolutely had to avoid was sliding backwards. Back in the States on concrete runways or ramps, we often "backed up" the aircraft using reverse thrust on the propellers. If you were to do that when on skis, the tail of the skis might dig into the snow and the aircraft would tip over backwards onto its tail. All in all, however, landing and taxiing on skis wasn't very difficult.

Taking off on skis is an art. Most people riding on an airliner are used to a takeoff as being a smooth continuous acceleration along the runway for about 30 seconds, followed by the nose of the aircraft rising up and finally lifting from the runway.

When you're trying to take off on skis, however, the friction of the skis sliding through the snow might be so great that you never reach the speed needed to raise the nose. Especially when the snow is deep and soft, aircraft skis will "plow" through the snow, as opposed to sliding along the surface. Even with your engines at maximum power, you might only reach 40 to 50 knots—well below the necessary rotate speed of 60 knots (roughly 70 mph) needed to lift the nose off the snow. And, yes, I said 60 knots; about half the speed the pilot of a large aircraft is accustomed to using. In soft snow, you need to raise the nose ski off the snow to allow the aircraft to accelerate, but if you're plowing through the snow at less than 60 knots and you pull back on the aircraft's yoke to lift the nose, quite often nothing will happen. The nose of the aircraft will remain right where it is and the aircraft will continue to slide along the skiway but not accelerate.

The solution is to do the opposite of what every pilot is trained to do. Instead of pulling back on the aircraft's control, you must push the yoke down and do it very rapidly. Then, as the nose of the aircraft drops, pull the yoke back into your lap as hard and fast as you can. In effect, you "bounce" the nose into the air using the compression of the nose landing gear strut as a spring. With the nose of the aircraft slightly pitched up, air flows under and over the wing at just the right angle to develop lift. The faster you go, the more lift the wings generate. The 22-foot long main skis no longer sink so deeply into the snow. At about 85 to 90 knots, your 60-ton airplane will lift from the surface, but you aren't looking at the gauges on the instrument panel. You're looking outside the aircraft, ensuring the aircraft's nose doesn't drop. You know the skis have lifted from the snowy surface because you no longer hear the sound of the skis smashing through ridges of snow, but now you have another problem.

You're moving so slow that if an engine fails, you won't be able to maintain directional control. Let me put that in plain language—not in pilot jargon. It means that no matter how much you push the rudder pedals or turn the yoke ("steering wheel"), the aircraft will start corkscrewing sideways in the direction of the failed engine. Eventually, a wing will hit the snow and you'll cartwheel along the surface, rupturing the fuel tanks and starting a fire. You'd be lucky to crawl out of that alive. If an engine were to fail at that slow a speed, your only practical option would be to pull back power on all the remaining engines and crash straight ahead.

As the aircraft initially lifts off the surface of the snow at a relatively low speed, the top priority is to accelerate the aircraft quickly to a speed of at least 120 knots or so. To do that, you must overcome another ingrained habit, namely holding the nose of the aircraft up. When you see an airliner taking off, the aircraft is climbing at an angle of 15 to 20 degrees. When taking off on skis, however, you gently ease the nose down and fly level as near to the surface as possible—perhaps 20 feet above the snow—to allow the aircraft to accelerate. Pilots will recognize this as staying in ground effect. Once again, you aren't looking inside the aircraft at the instrument panel; you and the copilot will determine altitude by looking outside. The flight engineer will look inside at the airspeed indicator and call your airspeeds out loud. As the aircraft accelerates past 120 knots (the normal rotate speed), you gently ease the nose up and climb out. The Hercules is a large aircraft so if you pull back on the yoke too abruptly, you could cause the aircraft's tail to hit the ground. Any turns at this low an altitude would cause the wing tip to hit the surface, cart-wheeling the aircraft across the snow. This is real "seat of the pants" flying.

A ski takeoff, even from a well-maintained surface, is by no means a sure thing. At McMurdo and the South Pole, Navy Seabees—the navy's civil engineer force—and NSF contract engineers would grade the skiway surfaces relatively smooth using road-building equipment. But if the snow is too soft, your aircraft can slide through the snow for miles, never gaining enough speed to get airborne. Like so many other things you learn from experience, if you can't get airborne on your first attempt, you might be able to take off on the second attempt if you can keep the skis in exactly the same tracks as the first takeoff. It also helps to wait ten to 15 minutes for the snow that was compressed on your first takeoff attempt to harden. I don't know why the snow hardens, but our experience was that it did. Second takeoff attempts usually worked. If a third attempt was necessary—and it didn't work—the flight would be cancelled. There was one more take off procedure we could use if the snow was too soft—JATO—which stands for Jet Assisted Take Off. Each JATO bottle was about the size of a golf bag. Four bottles were attached to each side of the aircraft, low on the fuselage just behind the wing. When all eight bottles were ignited, the total additional thrust was equivalent to one of our engines. Using JATO was like having an extra engine but it only lasted for just under a minute, long enough to get airborne but not much longer. I only used JATO twice in three years.

You had to be careful when using JATO. Many years before I flew in Antarctica, one of our C-130s made a JATO takeoff from an open-field landing site that had an extremely rough surface. The shock of impacting these ridges of ice during takeoff caused a JATO bottle to break loose from its attachment on the fuselage. Like any rocket, it flew straight ahead at great speed. Unfortunately, it hit one of the propellers, which broke loose and hit the adjacent engine. The Hercules had barely gotten airborne, and with two engines failed on one side, at a very slow airspeed, the aircraft crashed. Luckily, no one was injured and the crew was rescued. Nearly 20 years later, the aircraft was finally recovered, although another aircraft was lost in the process because of pilot error—with a loss of two lives.

Flying the Huey Helicopter

It was as a helicopter pilot that I became most directly associated with scientists and their projects. Helicopter flying in Antarctica was a world apart from flying the C-130 Hercules. Unlike the C-130s, only a few of the Huey flights were strictly cargo. Nearly all the helicopter flying in Antarctica was in direct support of National Science Foundation scientists, who conducted fascinating research through a broad spectrum of science disciplines. Helicopter pilots and crewmembers became good friends with many of the scientists, and we went well beyond what was expected of us in setting up camps and flying in supplies.

The UH-1N is best known as the Huey. Our Hueys weren't modified for Antarctic flying. Unlike the Hercules, for example, we didn't need skis. The Huey's skids worked just fine on any flat surface. One limitation we did impose on ourselves, however, was to fly only in good weather. In the days before the GPS satellite based navigation, flying in the clouds in mountainous terrain was suicidal. Another nice difference between the Hercules flight operations and those of the Hueys was that the helicopters had a hangar, and it was right in Mac Town. We could walk to work. While all the maintenance for the C-130s was done in the open, the far smaller Hueys could be pulled inside the helo hangar in Mac Town, out of the wind and cold.

One of the missions for our helicopters was taking scientists to the numerous penguin rookeries on Ross Island, all of which were located well away from McMurdo Station. On Ross Island, most penguins were the Emperor and Adelie species. The aptly named Emperors stood about three feet tall and their black and white feathers were trimmed with golden plumage. The smaller, plump, and noisy Adelie penguins were named by a French explorer after his wife Adele. (Unimpressed by the honor, she divorced him soon after his return to France.)

The explorer Ernest Shackleton established the base camp for his 1906 expedition near one of the largest Adelie rookeries on Ross Island. Shackleton could not have chosen a more beautiful site. The sheer black cliffs of the Transantarctic Mountains lie directly across McMurdo Sound, which teems with killer whales. Only a few hundred feet from the camp were thousands of the squawking, comical Adelie penguins. Behind Shackleton's campsite rises Mount Erebus, a live volcano that has been continually pluming smoke for centuries. The entire setting, like much of Antarctica, was a photographer's dream, and even a rank amateur like me wound up with one of his pictures published in the Smithsonian Institution's Air and Space magazine.

We often flew to the various early Antarctic explorers' base camps, located several miles apart on the shore of Ross Island. After a century, the buildings remain pristine because of Antarctica's exceptionally low humidity and complete lack of insects. Period newspaper and personal items inside the main buildings look as though the occupants had only just left and would soon return. The anchor in front of Shackleton's hut, used to moor the expedition's ship to the shore, still lies embedded in the sand in front of the main building. I felt like I was entering a shrine when going inside those buildings.

Another of our helicopter missions was to fly oceanographers to the very edge of the pack ice for them to collect samples for their investigations. The primary hazard in landing near the edge of the pack ice was the strength of the ice itself. There was no way to know how much weight the pack ice could bear. The strength depended on the thickness and temperature of the ice, neither of which could be determined prior to landing. To play it safe, we typically landed several hundred yards from the ice edge.

After loading supplies onto sleds, we hauled the equipment on foot to the ice edge. Often, we'd stay at the ice edge while the scientists completed their work. On one of these trips, we cut a hole about three feet square through the ice so the scientists could take water samples. The reward for us was that after the samples were taken, we had a perfect view of the sea life in the bright, clear water below. The Adelie penguins, who waddled so awkwardly on the surface of the ice, could now be seen as the most graceful of swimmers. They streaked and darted after small fish, changing direction and depth in a split second to catch their prey. They would then reverse course, zoom to the surface, and land upright near us at the ice pack's edge. Curious and unafraid, the Adelies would waddle up to us in a group before moving further down the ice's edge where they'd jump back into the ocean. One of the scientists, on his knees and still wearing his diving gear, waddled and squawked along with them. The Adelies were neither fooled nor scared and simply waddled past him.

Despite the amusement in watching penguins, you had to be alert for leopard seals and killer whales. Both were carnivorous and both hunted penguins. If we saw whales or seals, we always moved well away from the ice edge. Penguins instinctively knew the danger from killer whales. Before jumping in the sea, the penguins would mass up and peer over the edge of the ice into the sea, looking for killer whales or leopard seals. When the first penguin jumped (or was pushed) into the water, the others would watch to see if he was attacked. If not, the entire group would quickly jump in.

Staying out of the water and on top of the ice, however, didn't guarantee safety for either humans or penguins. The killer whales cruised below the ice near the edge, looking through the translucent ice for dark objects above. When a dark object was seen, the whale zoomed straight up, impacting the ice with its blunt nose. The ice would shatter and the prey was thrown into the air. Typically, the whale caught the penguin in midair before it fell back into the water. There was no way for the whale to know just what kind of animal it was hunting when it saw the dark shape from below the ice. Personally, I know of no whale attacks on humans but about a year after I left Antarctica, a leopard seal attacked a member of a private expedition.

Working at the ice edge posed one other amusing predicament. Not only are penguins unafraid of humans, they are also very curious about us. Frequently, when returning to our helicopter, we'd find our aircraft completely surrounded by penguins. There was a law that prohibited any kind of harassment of penguins, so we had to walk very slowly to our aircraft, slide open the door, and quickly get aboard before they followed us in. They were completely unafraid. It was only when we started our engines and the rotor blades began turning that they would waddle away.

Much of our mountain flying in helicopters was to the nearby volcano, Mount Erebus. Our Hueys had two engines whose extra power gave us the ability to fly higher than the single-engine Huey. Nonetheless, we were still limited in what we could do at the 13,000 feet elevation. We could orbit around Mount Erebus' caldera for scientists to take readings and measurements, but we couldn't land. Landing or hovering a helicopter requires far more engine power than simply flying through the air—more power than we could develop at high altitude. The thin air also meant we had to wear oxygen masks, an unusual event for the Huey helicopters, which are unpressurized and typically fly at low altitude.

Mount Erebus had one additional hazard—it was a very active volcano. As you flew around the crater's rim, you could clearly see red molten lava at the bottom of the caldera that sporadically ejected volcanic bombs. The crater rim and the icy slopes were littered with this volcanic debris. Most were only a few inches across but some were the size of small cars. Since any of these could easily damage an orbiting helicopter, we stayed away from the crater whenever we saw Erebus was throwing bombs into the air.

Mountain flying in helicopters is always hazardous. One of our helicopters crashed along a high altitude ridge in the Transantarctic Mountains but by good fortune, the two pilots aboard were only slightly injured. The aircraft had circled over a ridge, out over the valley, and back over the ridgeline when it started losing altitude. Even with its engines at maximum power, the helicopter hit the ice hard, splaying its skids, and rolling on its side. The rotor blades snapped as the helicopter rolled over. Luckily, there was no fire. Like dozens of other aircraft over the decades of flying in Antarctica, its wreckage remains where it crashed.

The South Pole

By mid-November of each year, the temperature at the South Pole finally warmed up to -65 degrees Fahrenheit, and we could start flying people and supplies there. The C-130 aircraft had been designed by engineers at Lockheed Aircraft Corporation to be flown in temperatures as cold as -65 degrees Fahrenheit, a standard design feature for most aircraft I've flown. If the temperature were colder than that, there was no guarantee that the structure of the aircraft would be safe. For example, fuel lines to the engines might start leaking or our hydraulically operated flight controls might malfunction. Even worse, for about nine months out of the year, the air temperature was so cold at the South Pole that the fuel in your tanks would freeze.

The first flight to the South Pole each November would be the first contact the South Pole crew of 20 people would have with other humans in nine months. Flights to the Pole were breathtaking. For the first half of the flight, you flew over the Ross Ice Shelf, with the 10,000-foot high Transantarctic Mountains to your right. Enormous rivers of ice oozed slowly through the mountain valleys, many miles wide and up to a hundred miles long. The glaciers were often heavily crevassed, sometimes at right angles to each other, forming hundreds of square-sided towers of ice cascading down the valley. This was the route of the early Antarctic explorers, such as Captain Robert Falcon Scott of the Royal Navy and Sir Ernest Shackleton, on their grueling attempts to reach the South Pole early in the twentieth century. As we flew over the mountains, the Antarctic Ice Sheet stretched ahead without limit—flat, featureless, devoid of life. It created the impression we were much closer to the surface of the earth, because the ice cap was up to 10,000 feet high and we only flew at about 25,000 feet.

South Pole Station was difficult to see. Much of the main building was covered by snow, and only a few towers and antennas poked up. The South Pole Station was at an elevation of about 9,300 feet and the skiway was about two miles long.

Most of the cargo we supplied to the South Pole was fuel oil to run their station. After landing, we would ski the aircraft near the entrance of the hemispheric dome of South Pole Station, where we would connect fuel hoses from the aircraft to fuel tanks on the ice, which were the same shapeless bladders used at Byrd Surface Camp. We carried a large fuel tank in the cargo compartment, and we pumped its contents, along with whatever we could spare from the aircraft's fuel tanks, into the South Pole Station fuel bladders. We lowered the aft cargo ramp to off-load supplies. During the entire time, we kept all four engines running because of the possibility of not being able to restart an engine, but the low temperature, the lack of oxygen at the high altitude, and the engine exhaust fumes could be nauseating. I often needed to put on an oxygen mask while we were on the ground to keep from getting hypoxia.

Nearby was the ceremonial South Pole, a silvery globe mounted on top a striped pole. Flags of nations signatory to the Antarctic Treaty encircled the ceremonial pole. Although the ceremonial South Pole was not precisely at the geographic South Pole, due to the creep of the Antarctic Ice Cap, the shiny sphere was nonetheless one of the most photographed sites on the continent—the ultimate "been there, done that"' photo. Just for fun, I took my cross-country skis with me on one flight, put them on, and skied around the pole, which meant I had skied around the world, right?

Whiteout Landings

In Antarctica, the weather is notoriously difficult to forecast. With only a handful of weather reporting stations spread over a continent larger than the United States, our meteorologists had too little data to predict what would happen, though these meteorologists were arguably among the best in the navy. It was often up to us in the aircraft to spot trends in the weather as we were flying. Weather observers at McMurdo, South Pole, and Byrd reported the weather at least once an hour over the high-frequency radios, so we could see immediately if there was deteriorating weather.

On one of my C-130 flights from Siple Station to Byrd Surface Camp I was caught by one of those unforecast Antarctic storms. The weather was deteriorating, but I had no option other than to proceed to Byrd. I didn't have enough fuel to go anywhere else.

The storm resulted from katabatic winds—the downslope winds in Antarctica which produce some of the strongest surface winds in the world, reaching peak velocities of nearly 200 miles per hour. There was no way to predict these winds although the reason for them was well understood. The continent-sized Antarctic Ice Sheet is nearly 10,000 feet at its highest elevation. The air directly over the ice becomes supercooled and consequently heavier than the air above. Under the force of gravity, this heavy air flows and accelerates down the long slope of the Antarctic ice sheet for hundreds of miles towards the ocean. As the winds scour the surface of the snow, ice crystals and snow are picked up, resulting in extremely low visibilities.

Katabatic winds flow over and around the surface terrain, resulting not just in zero visibility but very strong updrafts, downdrafts, and crosswinds. The resultant windshear and turbulence could put you at the limit of being able to control your aircraft. The winds stay relatively low in the atmosphere, at altitudes less than a few thousand feet, and typically last only a few hours.

As I continued my flight to Byrd Camp, the weather reports showed the wind speed continuing to increase and the visibility deteriorating. Finally, the visibility was reported as zero feet with winds of 40 miles per hour. Back in the United States, I would have diverted to another airfield. The problem in Antarctica is that for all intents and purposes, there aren't any. And I was running out of fuel.

Our operations manual listed an area near the Byrd Camp that was supposed to be relatively smooth and free of crevasses that would be suitable for a zero-zero landing. Our navigator used his radar as I prepared the aircraft for landing. The flaps and skis were extended as I slowed the aircraft to 120 knots (140 mph). We headed directly into the wind and started a gradual descent. Most airliners will descend at 700 to 800 feet per minute when on approach to landing on a runway and then smoothly raise the nose just prior to touchdown, but I wouldn't be able to see the surface until we hit it. Impacting the ice at 800 feet per minute could damage the aircraft, so I had to fly at the much slower descent rate of 200 feet per minute. I had to maintain that rate of descent until we impacted the ice.

I used my flight engineer to adjust the four engine throttles, telling what power to set, so I could concentrate on looking at the aircraft's primary flying instruments: the altimeter, compass heading, airspeed, and rate of descent. The copilot and navigator also called the airspeed and altitude out loud from their own instruments as a cross-check. With the flaps, landing gear, and skis extended, I descended into the swirling void. I looked out the cockpit windows to my left, but the snow was so thick I couldn't see the wingtips only 50 feet away.

The plane lurched and twisted in the turbulent air, but I was able to control it. As I neared the surface of the ice, I saw something out of the corner of my eye through the windows down by my feet. It was crazy. Snow appeared to be blowing from back to front, as though we were going backwards. I quickly looked back at my own instruments. My airspeed, altitude, and compass heading were normal, plus the call-outs from the copilot and navigator, using completely independent navigation systems, agreed exactly with my flight instruments.

I continued descending the aircraft because I trusted my aircraft's instruments far more than I did what I had just seen. Antarctica is known for its deadly visual illusions, although I had never heard of this one. Nonetheless, the conflict between what had to be an optical illusion and what my airspeed, heading, and attitude indicators told me was disorienting. For one of the few times in my life as a pilot, I felt my skin flush and pulse speed up. Had the temperature not been so very cold, I'm sure I would have been sweating. Only half a minute later we touched down on the ice, wings level and straight into the wind, just as we should have been. To this day, I have no idea how the snow could have appeared to be blowing from behind us, but there wasn't much time to think about it. I had a bigger problem. I had to find Byrd Camp.

Because of our slow descent, I landed many miles away from the field. As I turned the aircraft toward Byrd Camp, very strong gusting winds struck the aircraft from the rear, causing the plane to veer left, and then jerk to the right. The problem I faced is that aircraft are like old-fashioned weathervanes—they want to turn their noses into the direction of the wind. For the C-130, the tendency to weathervane was worse than any aircraft I've flown before or since, the reason being that the C-130 has a relatively large tail. When in flight, the tail of an aircraft is designed to keep the aircraft going straight, like the feathers of an arrow. On the ground, however, strong winds blowing on the tail from behind the aircraft, or nearly so, make the nose of the aircraft want to turn back around into the direction of the wind, like a weathervane.

Like most large aircraft, the C-130 has a steerable nose wheel, controlled by a small steering wheel in the cockpit. Since the nose ski was attached directly to the nose landing gear, I was able to turn the nose ski, but with the strong gusting winds and slippery icy surface, the nose ski kept sliding sideways.

I had to use what pilots call "differential power." When the strong, gusting winds twisted the nose of the aircraft to the left, I quickly added more power to the engines on the left wing than on the right, making the aircraft want to twist to the right. When the nose of the aircraft lurched to the right, I added more power to the engines on the right. Because the wind's direction kept changing, I was continuously adding or pulling power on the engines.

Even worse, the wings of the C-130 tend to rock up and down when you taxi over an uneven surface. The wheels and skis on the Herc are mounted on the left and right side of the fuselage, relatively close together compared to the C-130's very wide wingspan. The only way to reduce this rocking is to taxi slowly, but even that might not be enough. I fought the winds for nearly an hour as I taxied over the undulating snow before reaching Byrd Camp. The visibility was still so poor that the navigator had to guide me to the camp using his radar until we were within only a few hundred feet. When the campsite was finally in sight, we stopped the aircraft and opened the main door. We kept the engines running and several crewmembers went outside to hook up the fuel hoses.

I had one more lesson to learn. When I went to the main door to leave the aircraft, the surface of the snow was a fuzzy, milky white. I looked out and saw one of the small sheds at the camp but it seemed to be floating in a white void. I sat on the steps of the boarding ladder until my foot finally located the surface. By focusing on my boots, my eyes were able to focus on the snow they were resting on. What I was experiencing now was whiteout. More people have died in Antarctic aircraft accidents because of whiteout than any other reason.

In 1979 an Air New Zealand DC-10 aircraft on a sightseeing flight slammed into Mount Erebus near McMurdo, killing all 257 people on board. The cause of the accident was eventually determined to be the visual phenomenon called whiteout. The term whiteout is unfortunately (but correctly) used to describe two entirely different visual effects, and I had experienced both in my whiteout landing at Byrd Surface Camp. To most people, whiteout means that the snow is blowing so heavily that you cannot see very far. In that context, whiteout means your visibility is reduced to zero and the clouds above are so low they are considered to be zero feet above the ground. I will refer to this kind of whiteout as zero-zero weather.

The other whiteout—the one that can kill you—occurs in clear air. You can see mountains or buildings from many miles away but the surface of the ice becomes a fuzzy flat surface, even though the ice-covered terrain may be hilly. Sometimes, objects appear to float in space, such as I experienced when arriving at Byrd Camp.

This type of whiteout occurs when you and your aircraft are flying above an ice-covered surface and below a solid layer of clouds above. Sunlight diffuses as it passes through the cloud and hits the snow-covered surface below at many different angles. This light is then reflected off the snowy white, upward to the ice crystals that form the overcast above. This light is in turn reflected back down from the icy cloud at multiple angles to the snowy surface below and back up to the overcast and back down to the surface ad infinitum. The effect of this continuous reflection of light between the surface and the clouds is to completely wash out all shadows. There is no texture to the surface, no contrasting shadows that define changes in the terrain.

There's another deadly illusion in whiteouts. In mountainous areas, instead of seeing a slope that is rising up and meeting the overcast clouds above, you may see a level surface below that seemingly stretches out to infinity, creating a false horizon. There is no visual indication of rising terrain ahead. That's what happened to the Air New Zealand crew.

The danger of whiteout is clearly a serious and deadly issue when flying in Antarctica—or any other snow-covered terrain such as Alaska and the Canadian Arctic. To help identify the potential for whiteout, routine weather reports in Antarctica include surface and horizon definition. GOOD surface definition is a clearly visible surface on which you can see hills, slopes, or any other textural contrasts. The scale continues down through FAIR, POOR, and NIL.

Yet, surface and horizon reports might not be enough to alert you to the danger of whiteout, because the whiteout could occur in only one sector of the sky—commonly called sector whiteout. You might see mountains over 50 miles away off to the side, while the sector ahead of your aircraft might be NIL and NIL. You might be flying straight toward a mountain—like the Air New Zealand crew—all the time thinking you're over level terrain.

Despite all our precautions, one of my squadron's C-130s experienced sector whiteout while flying low over an open field-landing site. The pilot inadvertently flew into the side of a low hill while circling to land. Luckily—very luckily—the skis were fully extended for landing and the slope of the hill was quite gradual. Nonetheless, the impact was hard and the aircraft bounced roughly back into the air. The landing was abandoned and the aircraft returned to McMurdo with only minor damage. The Herc is one tough airplane.

Open Field Landings

For us C-130 pilots, the riskiest—and most exciting—flights were those requiring an open field landing, in which we took scientists to remote locations other than an established base. Most of the locations had been photographed, but I clearly recall several charts derived from these photographs that were mostly blank except for the wording in big bold letters stating that terrain elevation was not accurate. From old black-and-white photographs and charts, we tried to determine if the surface of the glacial ice was free of crevasses, and if in mountainous areas, whether there was enough room to maneuver the aircraft in the valleys. The final factor—the weather—would not be known until the day of the proposed landing itself. The sky had to be absolutely clear to ensure we wouldn't inadvertently fly into a whiteout.

For science teams going to remote locations, we typically paired scientists with specific crews so the pilots and navigators would know where the scientists wanted to go and how much gear they would be taking. The scientists would, in turn, learn of the feasibility of various landing sites.

Strong Antarctic winds often sculpted the surface into a series of more or less parallel ridges of hard snow and ice, called sastrugi, which could be several feet high and many miles long. From the air, sastrugi resembled waves in the ocean. Landing an aircraft on sastrugi would be like driving a pickup truck across a plowed field at 140 miles per hour. Some of our aircraft had been seriously damaged when trying to land on sastrugi that were too high, too close together, or too hard.

Most of the open field landings I made were to place science teams on site, but some open field landings were made solely to preposition supplies, such as fuel, for future use to refuel helicopters. One such site was the Lhasa Nunatak, about 300 miles from the South Pole in the Transantarctic Mountains. A nunatak is the tip of a mountain that has been almost completely covered by an ice sheet.

There were no scientists to tell us about the nature of the glacial ice, such as whether it was blue ice, crevassed, or deep soft snow requiring JATO to blast our way off the ground. On the charts, Lhasa Nunatak was a nine-mile long mountain ridge, located on the south side of the Transantarctic Mountains.

From the air, Lhasa Nunatak was easy to find since it was near several other mountains I was familiar with. I descended my C-130 to a thousand feet above the ice and made a visual inspection of the landing surface. In the clear Antarctic air, the sun reflected brightly on the glacial surface. The texture of the snow looked excellent. The surface appeared relatively smooth, free of crevasses, and snowy white, but the visual inspection was no guarantee. The only way you were relatively sure there were no crevasses was to fly a "ski drag" across the snow's surface.

A ski drag is nothing more than skimming the surface of the snow with the tail end of the main skis, putting only part of the weight of the aircraft onto the snow itself. If there are crevasses below, the force of the skis on the snow should cause the snow bridge to collapse.

I picked a site where I could fly the ski drag into the wind, enabling me to drag my skis along the surface for about two miles. The length of the ski drag was supposed to be no less than the distance required to land the aircraft plus the distance to take off, but there were no mathematical formulas or charts that could give you those distances. The weight-bearing capacity of snow and its coefficient of friction vary widely with the temperature and crystalline structure of the snow and ice. Dragging for about 60 seconds always gave me the landing distance I needed.

To fly a straight line, I picked the highest peak on the Lhasa Nunatak ridgeline and aimed for it. I slowed the aircraft, lowered my flaps and skis, and started descending. Snow surfaces are rarely smooth, and as soon as the main skis touched the snow I felt shuddering jolts as the skis the aircraft impacted the ridges of snow. You could tell a lot about the snow from the sound of the skis. If there was very little sound, the snow was soft. Sometimes, the snow was so soft that the friction of the skis through the snow would slow the aircraft. The snow at Lhasa Nunatak was almost perfect—firm and relatively smooth. After dragging along the surface for about one minute, I added maximum power to my engines and climbed the aircraft off the ice. Turning left, I flew a racetrack pattern alongside the ski track to see if I had collapsed any snow bridges. Everything looked good so I returned for landing.

I needed to land exactly in the tracks I had dragged through the snow. The reason: it was entirely possible that my ski drag was parallel and very close to a hidden crevasse. It was also possible that crevasses were located just before or just after the ski tracks. There were many bets and much joking made among crewmembers as to how closely the pilot could land in his own tracks, but it was deadly serious business. For us navy pilots, the closest analogy was landing on an aircraft carrier. The margin for error was about the same.

I landed in the ski drag tracks and stopped as quickly as possible. The loadmaster lowered the aft ramp and cargo door, and we quickly unloaded the fuel pumps, hoses, and the large fuel bladder. It was an all-hands effort and rank was irrelevant. I was the squadron commanding officer, but the loadmaster was in charge of this part of our operation.

Everyone pitched in and did what the loadmaster told him to. It was actually a relief to unstrap from the pilot's seat and do hard manual labor.

My flight engineers hooked up the fuel hoses from the aircraft to the fuel bladder, and we started to fill it. The loadmaster marked the landing site with radar reflectors and flags to help our helicopters find the site. When the offloading was completed, we closed up the aircraft's aft ramp and doors and prepared for takeoff. The snow was firm, and we took off well within the distance of the ski drag remaining.

Exploring the Transantarctic Mountains

Several of the open field landing sites were built up afterwards as seasonal bases, such as the ones near the Ice Streams on the Ross Ice Shelf and on the Beardmore Glacier. Tent cities were erected, providing comfortable shelter for sleeping, a galley, and even limited laboratory space. After the camps were set up, we would receive weather reports each hour, including the critical surface and horizon definition. After the first open field landing, we didn't have to perform the ski drag each time prior to landing provided we landed on the same ski tracks.

I returned to Lhasa Nunatak refueling several weeks later, flying in one of my Hueys based out of the Beardmore Glacier Camp. We were flying back from re-supplying some geologists high in the Transantarctic Mountains. I greatly enjoyed the Beardmore Camp. I talked with the scientists during chow about their fascinating projects, although with so many scientists from Ohio State, football often dominated the conversations. The atmosphere was egalitarian: PhDs, pilots, and mechanics all took turns washing dishes and cleaning tables. When there was time off, some people played volleyball on the ice, but my passion was cross-country skiing.

Flying helicopters at Beardmore enabled me to scout the surrounding areas for safe places to ski. I glided almost effortlessly on my cross-country skis for miles across the glacier, far from the sounds and sight of our camp. In my jacket, I carried a survival radio along with an extra battery. I stowed ice-climbing crampons in my backpack and strapped my ice axes to the outside of the pack. As I skied up some of the slopes, I could see up the glacier to the seemingly endless Polar Ice Cap and down the Beardmore as it flowed like a river of snow and ice through the Transantarctic Mountains to the Ross Ice Shelf below. I have never experienced such solitude and absolute stillness in my entire life as when I skied miles away and out of sight of the camp. Few people, if any, had ever been where I stood. So vast is Antarctica that it's likely I was the first human at several of the sites we flew to, and I think for some of us that was an irresistible draw in choosing to go to Antarctica. Where else on this planet can you go and quite literally be where no human being has ever set foot before?

Rescue At Sea

My tour of duty in Antarctica overlapped a privately funded expedition in the mid-1980s called "The Footsteps of Scott." The expedition was organized in honor of the legendary but ill-fated 1912 expedition to the South Pole by Captain Robert Falcon Scott of the British Royal Navy to the South Pole. The members of the Footsteps of Scott expedition were not novices. Most were former members of the British Antarctic Survey (BAS), the United Kingdom's Antarctic program, including my friend Giles Kershaw, the expedition's pilot.

The very day the Footsteps of Scott expedition reached the South Pole, the expedition's ship, Southern Quest, became trapped in pack ice off Ross Island and was in danger of sinking. This happened in January, at the beginning of the Antarctic summer when the surrounding frozen sea ice was starting to break up. Unlike the Arctic Ocean, which retains a large area of frozen ocean throughout the year, Antarctic sea ice largely goes away each year. This breaking up of the ice is as much the result of the ice drifting away as it is of melting.

The frozen sea ice adjacent to McMurdo, however, remained in place for many miles between the ship's pier at McMurdo and the open sea. The only reliable way to get our oil tanker and cargo ship to McMurdo each year was to use an icebreaker. One of the US Coast Guard's icebreakers would break its way through the ten-foot thick frozen sea, clearing a path for the two supply ships.

The sea ice furthest away from the continent where the water is warmer breaks up first. The ice is thinner and more subject to fracturing than the thicker ice close to shore. Then, the wind and ocean currents move the ice away, drifting to the north where it finally melts. As the summer progressed, we in the Hercules, flying miles overhead, could see the full extent of the frozen ocean as it broke into vast sheets of flat ice, called ice floes. From the air, the sea looked like giant sheets of Styrofoam, etched by the inky dark ocean.

The water between the floes was called a lead (rhymes with bead). The Footsteps' support ship, the Southern Quest, was navigating its way through the leads when the ice floes, instead of drifting further apart, started moving together, gripping the ship like a vise. It couldn't break free, and the pressure of the ice against the ship's hull slowly crushed it.

Within only a few minutes of being wedged between the floes, the hull was crushed and the ship sank. If there is any good news about being sunk by an ice floe, it's that the ice floe also provides a refuge from the sinking ship. It becomes your lifeboat. Despite the rapidity of the ship's sinking, all 21 people aboard escaped onto the ice floes, but were able to salvage only a few possessions.

As soon as the distress call came from the Southern Quest, we launched every helicopter we had. I was on SAR duty that night and within minutes, we started up our helo's engines and took off toward the site of the sinking. For Antarctica, it was unusual weather. The sun orbited the continent at a low angle above the horizon, but at this hour it was very low and the cloud layer was very thick, turning the sea ice surface a gloomy gray.

Luckily for the stranded crew of Southern Quest, the icebreaker Polar Star had two helicopters, and they were closer to the sinking than we were. The Coast Guard got to the survivors first and took charge of the rescue operation. As we approached the site of the sinking, the survivors looked insignificantly small in the vastness of the frozen ocean. At a distance, had we not known better, we would have thought the band of survivors was yet another small family of penguins. The priority was to get the survivors off the ice floe as soon as possible, so all our helicopters picked up a few survivors at a time and flew them to the nearest safe position, a barren rocky beach on a nearby island. After everyone was picked up from the ice floe, we picked up the survivors off the beach. Our group of five was unharmed, but one was shaking almost uncontrollably. We then flew them to the helicopter-landing pad at McMurdo.

My own flying career in Antarctica ended only a few weeks later when the summer season closed. I never returned. In the years since leaving Antarctica, only a few of my friends ever went back to The Ice. For the majority of us, it was a one-time adventure. Sadly, my friend Rob Hall died only a few years later while leading a mountain climbing expedition to Mount Everest, a story chronicled in Jon Krakauer's book *Into Thin Air*. Giles Kershaw was killed in an aircraft accident in Antarctica on yet another private expedition. As I was writing this piece, yet another ship was sunk off Antarctica by icebergs.

Years after I flew in Antarctica, I was at a graduate school in Washington, DC, doing research at the Library of Congress for a paper on US foreign policy in Antarctica. By chance, I looked at a map of Alexander Island west of the Antarctic Peninsula and was surprised to see a very familiar name—my own. Unknown to me, after I left Antarctica the National Science Foundation had named an ice-covered peninsula after me—Derocher Peninsula. Having a landmark named for an individual is an honor often accorded to ship captains and aircraft squadron commanders, among others. However undeserving I am of such an honor, it nonetheless amuses me to think that whatever else I may do in life, there will be at least one legacy—a pristine ice covered peninsula seldom seen except by penguins and sea birds.

Huey helicopter at the ice edge with Adelie penguins curiously moving towards the aircraft.

Captain Derocher kneeling at ice edge, with penguins waddling up to him.

Front end view of C-130 ski plane at Willy Field skiway.

Captain Derocher (right) holding cross country skis in front of his helicopter in TransAntarctic Mountains.

C-130 ski plane at Willy Field skiway; flag was held up after returning from last flight to the South Pole for the season.

ANY LANDING YOU CAN WALK AWAY FROM
by F. Victor Sullivan

The airplane's nose lifted, I pulled back on the stick, and we rose effortlessly into the air. I was elated until we reached about 500 feet, when the plane bobbled. It wasn't serious, just enough to surprise me, but as I reduced power, the nose dipped a little and I thought, *You idiot, what makes you think you're a test pilot?*

Then I remembered the instructions from Burt Rutan, my plane's designer. He had warned, "It will startle you on your first takeoff."

After that minor upset, N62MV climbed smoothly to 3,000 feet, and, as I slowed to let my chase plane catch up, I realized this was one of the proudest moments of my life. It took years of dreaming, planning, and frayed nerves; this was the payoff. After three years of nearly constant work to build her, I was flying my homebuilt VariEze for the first time.

She and I logged 40 more hours of flight before that dream turned from beauty and grace into horror.

After dreaming about it since my childhood in the 1930s, I finally, in 1967, got the chance to pursue my dream of flight. When I showed up for my first lesson, however, I was in for a shock. The man who came to meet me, holding out his hand for me to shake, had only one leg. How was he supposed to teach me to fly?

I soon learned that Riley Helms was one of the best flight instructors around. He hadn't taken to flight naturally. Like most young men in this region, he worked as a strip miner in the coalfields that gave Pittsburg, Kansas, its name. His brother learned to fly, and he tried to get Riley to go up with him when he passed his test.

"Not unless I can leave one leg on the ground!" Riley said. I can picture him shaking his head at the notion of taking to the air. His focus was on the ground.

Riley was in a mining accident some time later. He was lucky just to lose a leg, but any way you look at it, it was a life-changing event. He was no good to the mines anymore. His brother showed up at the hospital while Riley was pondering his uncertain fate.

"Well, Riley," the brother said. "You can leave one leg on the ground now, so as soon as you get out of here, we're going up!"

Riley couldn't turn down a challenge like that, so as soon as he could, the brothers took flight. The result? Riley loved it so much that he decided to learn to fly himself, eventually earning his commercial and instructor ratings. His only restriction was that he had to fly aircraft with less than 11 inches between the rudder pedals so that he could turn his foot sideways to control both of them.

One of the things that made Riley a good teacher was the way he had learned. He received his instructor's rating from an old barnstormer who had taught airmen in the war, and that meant he learned things the old way—you learned to get out of a stall by doing it, to get out of a spin by doing it, and to make a dead-stick emergency landing by doing it. You either learned it, or the lessons came to a sudden end!

I will never forget one day when Riley pulled the throttle, turned to me, and said, "Your engine just quit. Find a field."

Of course, I knew Riley would give me back the throttle at some point saying, "Okay, you could have done it if you had to." This was just a dry run, and we'd already made dozens of them. I set up my approach as we had practiced. I was on the final approach at 500 feet, and as I reached for the throttle, Riley put his hand on it.

"Your engine's dead," he said.

I continued the approach, and at 100 feet, I reached for the throttle again. His hand was still there. At about 50 feet, I exclaimed, "We're going to have to land!"

We were passing over the road into a hay field, and I could clearly see the barbed wire fence below.

"Yep," he said laconically. "Looks like you're gonna have to land."

That hay-field landing was a bit rough for the Cessna Aerobat, but we made it to the ground in one piece.

I turned to Riley and said, "I sure hope you know the farmer who owns this field!"

Riley grinned. "I've been trying to get you to pick this field for the last three lessons!" We laughed as he added, "Now you know you can do it."

I'm sure his lesson saved my neck more than once. I did end up doing an engine-out landing in my VariEze when the wrong fuel cap was placed on the fuselage tank. Without a vent in the cap, it created a vacuum and the engine quit as I was on a rather long downwind leg. I realized I wouldn't make the runway, and so I put her into a cornfield just across the hedgerow from the runway. The only damage was that the nose gear folded, and that was quickly repaired.

My kids grew up in small airports, and our dog, a dachshund, earned her wings as well. Our airplane, a Beech Musketeer 180 was, to us, like a station wagon to other families. One memorable family trip was to the Boundary Waters of Minnesota. The week canoeing Minnesota's 10,000 lakes was wonderful, but the flight there was pretty memorable as well.

I filed an instrument flight plan and flew through one of the worst thunderstorms I had yet encountered. After a harrowing time of it, I made an instrument landing in near minimal conditions. And there were other complications. My wife, Mary-Kate, gets airsick. You can imagine how that added to the drama! As any pilot can tell you, the most beautiful sight in the world is not the Minnesota lakes, the Rocky Mountains, or the most amazing sunset ever. It's the runway lights when you break out of the clouds at the end of an instrument approach at minimum altitude with very little visibility!

Our family of four, plus dog, spent lots of time flying together. Family vacations were special times, and we all loved to travel. My daughter, Olive, shares her views of these family flights elsewhere in this book.

When the kids were in high school, Mary-Kate and I decided to embark on a new adventure. We decided to build an airplane. We had been researching homebuilt aircraft, and at last we settled on the VariEze, designed by Burt Rutan. We saw Burt's first VariEze during a family flying trip at the Experimental Aircraft Association fly-in in Oshkosh, Wisconsin, in 1977. Rutan was an aeronautical engineer who went on to design and build Voyager, the first plane to fly around the world without refueling. He also created the first civilian manned spacecraft, the Virgin Galatic.

Rutan's idea was to create a simple, lightweight aircraft for daytime flight in clear weather. I liked his philosophy. We also liked the plane's striking canard design, but because we are serious craftsmen, many other practical features attracted us as well.

According to Rutan, the plane's canard structure was designed to reduce stall, and winglets performed as both vertical stabilizers and rudders to control yaw. In a canard, the main wing is in the back, and the horizontal stabilizer is in front. The engine is in back as well. Interestingly, this is the way the Wright Brothers designed their first plane. The VariEze was also spin-resistant, a feature which attracted the attention of NASA. The agency eventually built two VariEzes for flight test and wind tunnel research.

The model we selected was the original Eze. It employed a new airfoil developed by aerodynamics specialists in Scotland. Rutan began publishing his first plans in 1976, and the first homebuilts began to fly less than a year later. For that era, that was an astonishingly short timeframe from plans to flight.

With a wingspan of 22 and a half feet and weighing only 585 pounds empty, the plane could carry two people for 700 miles at about 180 miles per hour. Without a passenger, most VariEzes could climb to altitudes of 25,000 feet.

We thought our son, Mark, a tinkerer like his forefathers (and mothers!) would enjoy helping us. We hoped to encourage our daughter to join in as well. And we had plenty of room in our huge living room. Yes, we decided to build the plane in our living room. Other members of the Experimental Aircraft Association had to build their planes in the garage or the barn, or even rent a building. Our workshop was handy and warm in winter and air conditioned in the summer.

The plane arrived in boxes and crates, which were full of huge sheets of thick blue Styrofoam and rolls of silvery woven fiberglass cloth. We had jugs of epoxy and stacks of tools. We even created some of our own tools. One such tool was a heated cutting wire that we used to slice the foam into the correct shape for the parts of the plane. It didn't take long before the dining room table was shoved to one wall, the couch to another, and sawhorses and workbenches had taken over the whole place.

The key to the VariEze's construction was its composite structure, basically, layers of fiberglass built up around a foam core. We used the hot wire to cut the foam to shape according to templates provided with the plans. Once the shape was ready, we applied the fiberglass cloth and coated it with layers of epoxy resin until it was solid and held its shape. The composite had the advantage of providing great tensile strength at very light weight, and has since become a common construction method.

We built the bulkheads first. Construction of the wings came next. The fuselage of the plane was built similarly to the way the Stearman and other early aircraft were constructed. Wood pieces were notched into the bulkheads, but instead of bolts, glue, or welding—epoxy and fiberglass held our structure together. This skeleton was then covered with a different type of foam that we carved into the desired shape. As all of the airframe pieces were completed, we began the process of assembly. With every part attached, the wings extended into the dining area, the front canard wing was almost against the east wall, and the mounted engine was about six feet from the fireplace wall on the west.

The final touches were a rebuilt 80 horsepower Continental engine, and the propeller, custom carved by Frank Bresnick, a friend, fellow EAA member, and very talented pattern maker.

Despite all the dire predictions from friends, when the time came to remove the plane from the living room, we simply took off the canard and the left and right wings and rolled it out the back patio door, as planned. A suspicious person might even think the house had been designed for such construction!

One day, the fuselage was on the driveway in front of the house without its wings so I could spray on the black primer paint while the plane rested on the engine, the nose pointing straight up. It made spraying the bottom much easier. Two neighborhood children were walking by, and I heard one say to the other, "Look, I told you it wasn't a plane, it's a rocket!"

At last it was finished. We set out for Atkinson Municipal Airport with the wings and canard in the back of the truck and the plane in tow. Three years had passed since the crates and boxes of parts arrived—three years of loving labor to create an object of beauty and grace.

On the maiden flight I didn't even retract the nose gear. I just flew straight and level until Carl Lipscomb, flying chase in his Cessna 182, caught up to take photos. Then, I returned to the field and made a complete trial approach before attempting to set her down. She was smooth all the way, with an approach not unlike that of the Mooney 201 I had flown for more than a hundred hours.

The difference was that the Eze came in like a jet, nose high, and touched down on the main gear as nice as you please before smoothly dropping the nose gear. I was down.

I taxied back to the parking slot in front of the terminal building to celebrate with Mary-Kate and our friends. What a day!

That first flight took place on July 10, 1980. During the 40 hours of test flights required by the FAA, I did my flying with lots of landings at Pittsburg and a few at Joplin. Air work tested the Eze's handling and allowed me to get familiar with the various speeds. The airplane simply would not stall or spin. Instead of dropping a wing and going into a spin at about 45 miles per hour, it simply started losing altitude in a very nose-high attitude, retaining its normal landing attitude. Even a high-speed stall simply would not occur. The plane would hold the steep bank and begin to lose altitude, but it was never out of control.

When I reached the magic 40 hours of flight time in the plane, I made an appointment for the final inspection, which would lead to registration. In the late afternoon the day before the inspector was to arrive, Mary-Kate and I went out to check the plane over one last time. Our son's best friend, Jeff Olsen, was living with us while he attended Pittsburg State University, and he came along to help. We removed the canard to check the mounting and the nose landing gear mount and retracting system. This required disconnecting the push rod from the joystick to the elevator actuation system. We checked the battery connections, the lights, and the wing tip strobes as well as the radio gear.

Everything looked good. As we reassembled the parts and checked the tightness of the bolts, Mary-Kate said, "I think this bolt is a little loose." She pointed out the bolt that connected the joystick to the push rod system. The bolts to the joystick did not use nuts with a hole for safety wire, but rather employed a new all-steel lock-nut that didn't require safety wire. It did, however, put extreme pressure on the bolt threads. The system was used for most of the fastenings in the aircraft.

I checked it, but I was ready to go home and relax. This had been a tiring day for me. I'd just come from a frustrating meeting at the university where I was dean of the School of Technology, and I was pretty tense. I told Mary-Kate to forget about it. "If we tighten it any more, the controls will be too stiff." She shrugged, and we closed everything up.

I added, "The inspector will want to check the fuel system, so why don't I just take her up and burn away some of the fuel so we won't have to drain it tomorrow? I'll do a few touch-and-gos, and we'll put her to bed." We figured we had another 20 minutes of daylight, so it seemed like a good idea. It was a beautiful September day, clear air with unlimited visibility, the way it used to be in Kansas before air pollution was such a problem. Dusk was falling and a breeze was increasing from the east. It was a typical day, and we had no indications of the trouble ahead. I went through my usual pre-flight checks, fastened my seatbelt tightly, as usual, and moved onto the tarmac.

Reaching Runway Ten, I took off into a slightly gusty southeast crosswind. It's the shortest runway, and narrow at about 30 feet. It was paved, but the black asphalt had no markings, making it hard to see at dusk. In fact, today it's used only as a taxiway. Everything went well as I made my first trip around the pattern, but since the runway was a bit short it required a correct approach. The Vari-Eze usually lands at 60 miles per hour, but on my second landing I was a bit hot. I was also a bit farther down the runway than I liked, so I decided to go around one more time. As I moved back on the stick, I felt a slight bump.

Wonder what I hit? I figured it was something on the runway.

When I reached 800 feet above the ground, I attempted to lower the nose and to start my left turn for downwind, pushing the stick forward.

Nothing happened!

I shook it and pushed forward again, still climbing. The stick was disconnected from the elevator push rod. The push rod should have been connected by the very bolt that Mary-Kate had pointed out—the one I had disregarded. But I wasn't interested in doling out blame at the moment. For the first time in my life I experienced true panic. If you lose elevator control, the plane crashes—that is all there is to say about that! This is it, Victor, I thought. You are dead! I seemed to lose all sense of time. I don't know how long it was before I came to my senses, but it was probably just a few seconds.

Then, I realized that the plane was still climbing. With no elevator control, that was impossible. We should have been plummeting straight down, auguring into the ground, ending as a huge fireball. But, I was still alive. So I pushed the mike button and called, "Mayday! I have lost elevator control. Mayday!"

Mary-Kate answered the call from the terminal where she waited and watched, and I said, "Call the ambulance and the fire department!" There was no control tower at this airport.

Her voice came back over the mike. "Are you sure? No elevator control?" With her flight experience, she knew what the loss of the elevator control would mean. Even knowing that this probably was a fatal situation, she remained calm and level-headed. No wonder she's the rock of my life.

"I don't know how, but I have no elevator control and it's still climbing," I said. "I'm going up to 3,000 feet to see if I can figure some way to descend with control."

We were both surprisingly calm after my initial moment of blank panic. Or maybe it's not so surprising. She is the calmest person I've ever met, and we are always at our best when we're brainstorming. The fact that I was still in the air was amazing, but the question was how to get down. It was simple: I had a problem, and I was solving it, with her help.

Mary-Kate adds that another reason she seemed so calm is that she was trying to soothe Jeff, who was shaking from the tension.

We exchanged ideas and tried to figure how I could reach the push rod. No luck. The cockpit was too tight, and it was inside the right armrest. For the first time, I regretted that the fiberglass construction was so strong.

Mary-Kate called the emergency in, and we talked again.

There are two ways to descend in an airplane: First, normally, using the elevators; or second, by reducing engine power. Since the first option was off the table because I had no elevator control at all, I had to rely on the second. At 3,000 (2,100 feet above ground level), I started slowly dropping the airspeed, pulling back on the throttle while monitoring altitude. I gradually reduced throttle and carefully watched the airspeed drop, trying to determine the exact point at which I would start losing altitude. From there, I could calculate how to fly the approach to land. Talk about being a test pilot!

A normal approach was to maintain about 100 miles per hour until turning final, reducing to about 65 miles per hour for final, and landing at about 60. I was counting on the canard configuration: both the main wing on the rear of the plane, and the front wing, which had the elevator—on the VariEze it's really a flap—are lifting wings. In the usual aircraft configuration, the horizontal stabilizer in back pushes the tail down to hold the nose up. If you lose elevator control in a conventional aircraft, the tail rises, the nose drops and the plane will go straight down with no way to control it. I was now halfway around my pattern and about midpoint of downwind, approaching 100 miles per hour. As I reduced airspeed more, the canard would have less lift and the nose would lower. That was our theory anyway. It worked.

WOW! Did it work! At almost exactly 100 miles per hour, the nose dropped instantly, and we headed straight down. The ground was rising at an alarming rate. I was accelerating directly toward the ground, feeling nearly weightless. Movement and throttle control were difficult in this condition and my heart was hammering within my chest. "Now what?" I chided myself. If I added power, would the nose come back up, or would I just smash into the ground at a higher speed? A number of very serious questions raced through my mind, none of which had a satisfactory answer.

I remember clearly thinking, *Well, if you're going in, it won't make any difference how fast you go, you're going to die.* I gave it full throttle, and our speed toward the ground and my death increased very quickly. At 2,000 feet, though, she came out of the dive at 120 miles per hour and climbed again. I worked on controlling my breathing, tried to relax, and reported the climb to Mary-Kate, adding, "I'll try again."

I wanted to lose altitude, but not at this rate! Not in such an uncontrolled manner. I'd regained control over myself. Now, I needed to have control over my plane. I tried again, reducing power and waiting for a descent. It was better. The second time, I only lost about 800 feet, but that still wouldn't lead to a successful landing.

Riley had told me many years ago that any landing you walked away from was a good one. I dearly wanted a good landing that day. I needed to be able to control the rate of altitude loss so I could control the landing. I knew I'd be coming in hot and hard, but the difference between a good landing and a bad one would depend on how hot, and how hard.

As I was climbing again, I realized another way to change pitch attitude was to change the center of gravity of the plane—more discussion with Mary-Kate. There's very little room in a VariEze cockpit, but it was a possibility. So I repeated my throttle reduction procedure and in addition, I shifted my weight by leaning as far forward as I could. When the nose dropped, I quickly leaned as far back as I could and shoved the throttle forward. I only lost about a hundred feet. *Maybe I have an answer,* I told myself. I reported the process to Mary-Kate and started my next practice roller-coaster descent. This time I only lost 80 feet. Each practice descent was better than the last one. For the first time, I felt like I might survive this flight.

By now it was getting pretty dark and, of course, the wind was up some more. Runway Ten didn't have lights, but I could see headlights on the county road. I also realized that my purpose for being up there had been to burn off fuel—I was now running on the fuselage tank, which only held five gallons, and it was far from being full.

It was night, with no runway lights, and because of the crosswind, the other, longer runway wasn't an option. And I had no elevator control. What else could go wrong? Oh, yes! I was running out of fuel, which would mean the engine could die. With no power, I would have no altitude control at all and no chance of survival. A situation like this could ruin your whole day.

It was time for a decision! It was time for action! I had no option, but to go for it, especially as conditions were getting worse. This was, indeed, the moment of no return. Do or die, for certain. I pushed that last thought out of my mind and focused totally on the job at hand.

I called the field again and announced that I was making a seven-mile straight-in final and would continue my descent. I could see from their flashing red and blue lights that the emergency crews had arrived and were waiting at the terminal building. As I turned final and started my new landing procedures—throttle out, lean forward to shift my weight toward the nose to descend, throttle in full, and lean back quickly to hopefully slow the rate of decent—I realized Murphy was alive and well once more. Murphy's Law simply stated is that if anything can go wrong it will. And it did. A strong, gusty windshear was threatening to blow me sideways off the runway. The crosswind, huffing and puffing stronger across my flight path, now caused the plane to yaw into a left turn every time I did the descent maneuver. Now I had a corkscrew added to my roller-coaster process, but I was descending with a bit of control, losing only 50 feet per dip.

I was afraid to get under about a hundred miles per hour, clearly remembering the terrifying dive earlier. This was a fast approach to a short, dark runway, but I had no other choice. I realized I had forgotten about the speed brake, a flap on the belly that could be deployed to increase sink rate for short field landings. Sailplanes have similar devices, as do almost all jets. I tried that, but it was designed to be deployed at speeds under 80 miles per hour, and it had an automatic retraction system that would close it immediately in case the pilot initiated a go-around. So I continued on in my corkscrew. As I passed over the road just prior to the runway, I could clearly see that I would be coming in high as well as fast, so at what I judged to be about 80 feet above the ground and about a hundred feet from the runway's end, I pulled the throttle and the speed brake and held the speed brake on as hard as I could. I knew it was a risk, but I also knew it was my only choice if I didn't want to overshoot the runway. I expected to crunch the landing gear, but I figured if I lived through this adventure, that would be easy enough to fix. With no other option, I held the speed brake tightly. I was so low by this time that I knew I didn't have time for another corkscrew. I had to land. It was literally do or die—or both.

Well, it worked, but again, my success was a bit too much because the plane dropped almost straight down, not in a nose-low dive, but like a rock, flat. We struck the ground at about 80 miles per hour or so—and just short of the paved runway. The nose wheel caught the edge of the runway and folded with a crash that shook the plane. The main gear, designed like a very strong fiberglass bow, spread wide and threw us back into the air, and we dropped very hard again 100 feet further down the runway, just off to the left of the pavement. I quickly closed all of the switches and shut the fuel switch. I was down and awake.

One fear I had was fire upon landing, so as quickly as possible, I opened the canopy and climbed out. When I stood up, I had a horrible pain in my back, and after a couple of excruciating steps I had to sit. So I did, on the front edge of the main wing. I could see the fire truck on its way, its powerful engine roaring and its emergency lights flashing in the darkening evening. That was the last thing I remembered until I woke up in the ambulance with Mary-Kate at my side.

My first question was, "Did I break the prop?"

She smiled and said, "Yes, very badly." I next woke up on the table in the operating room with lots of faces in masks looking at me. I passed out again and my next thought was that my heart had stopped and they were trying to start it again. Then I was up on the ceiling looking down at myself lying on the table. I thought, *I have to start my heart; how do I do that?* Then again I was on the table and again up, looking down at the group bending over me. I didn't wake up again until in the morning.

I spent 11 days in the hospital and was fitted with a back brace. I hadn't broken any bones, but the second and third lumbar disks of my spine were compressed. The doctor reported I was a half-inch shorter than I had been, but that I should be fine after a long period of rest.

The next afternoon the FAA inspector arrived to interview me. After I told him what had happened, he said, "You must be wrong. If the elevator control had failed, you could not recover from that."

"Yes, I could, because I did it."

He was skeptical, but he went out to study the wreck and returned to report that yes, the elevator control was disconnected, but I had landed it. "I've been talking to records in Oklahoma City and as far as we can learn, you are the only pilot we know of who ever lived through losing total elevator control," he said. "It simply cannot be done!"

I explained the canard design and the reason I had been able to keep some measure of control.

He said, "I'm telling you, it's just impossible."

The canard, with its unique elevator construction, was the cause of my hard landing. It was also what saved my life. It turned out that the fault was not the lock nut, but that I had taken the canard off a number of times during the construction and flight testing stages and had worn the bolt down. Specs for the bolt said that it should be replaced after each removal. I had never seen that information, so I'd never replaced it.

In retrospect, I think holding the speed brake was a mistake because it made the plane descend so quickly. However, at the time, everything was happening so fast I didn't have time to react. Also, I wouldn't have been so stupid when I put the plane back together. I would have checked that loose bolt, for one thing. The lesson here is, don't fly airplanes of any kind when you're stressed or short of time. The old saying goes, "Live and learn." Here's your chance to learn from my experience, so you don't have to do it the hard way!

I flew about 600 hours after that hard landing. Now, I enjoy aviation with my grandson, Frank Abshire. We share the joys of flight, and he even started working toward his private pilot's license. As for me, I

Victor Sullivan with grandson.

had the good sense to quit flying at the age of 72, when I had only flown 12 hours that year. I still have my pilot's license, but I haven't bothered to renew my medical certification. But I still get a crick in my neck from looking at the sky.

Rutan's original N4EZ is now on display at the Smithsonian National Air and Space Museum in Washington, D.C. As for ours, all that remains are a few photos of my triumphant first flight—and the canard wing, hanging from the balcony of our living room, where N62MV was born.

What about my last landing in the VariEze? According to Riley's definition, it was a good landing: I walked away from it, even if I made it just a few steps.

Victor - VariEze takeoff.

Victor - In flight - VariEze.

Victor - First flight.

RELUCTANTLY AIRBORNE

by Olive L. Sullivan

My father's romance with flight started early on, and by the time I arrived, he was thoroughly infected with the airplane bug. My memory was of being packed in the back of the Beechcraft with the luggage, but in reality, Dad took his first flying lessons in the late 1960s, when I was about eight or nine.

Once he got his license, our four-member family (plus dog) no longer squeezed into the cramped quarters of the family car for road-trips. Instead, we squeezed into the even more cramped quarters of the Beechcraft with the incessant engine noise, smell of airplane fuel, lurching turbulence, and the memorable sound of my mother using her sick sack in the front seat. I recall spending hours on runways in tiny airports in even tinier towns, everything from dirt or grass strips to paved runways to the city airports catering to private planes. One of my favorite memories is the sight of our dachshund, Pfeiffer, standing on a small paved runway somewhere in Western Kansas, her ears streaming back in the constant wind, looking for all the world like she was waiting for takeoff clearance from the control tower.

One of my most terrifying memories is of flying in a thunderstorm of biblical proportions into the tiny airport at Ely, Minnesota, where we were to embark on a week-long canoe trip of the Boundary Waters ... if we lived. I recently found out that it was my dad's first instrument approach. I was horrified, because it was only my complete confidence that he knew what he was doing that kept me from being the first high-school student ever to drop dead of fright. The only thing that kept me from kissing the ground on landing was the threat of my older brother's scornful sneer.

Another child might have loved this lifestyle (except for the sick sacks), but I wasn't that child. I detested the smell of airplane fuel. A child with an artistic bent, the only interest I could muster up was that planes had a sort of sleek beauty about them.

One summer when I was in high school, a friend and I accompanied my parents to the annual Experimental Aircraft Association Fly-in in Oshkosh, Wisconsin. I went along for the promised canoe trip down the Wisconsin River; my friend went because, curse it, she loved to fly. She later joined the Air Force. Sigh. I spent the fly-in exploring the older barnstormer craft, which to me represented the real romance of flight, since I'm interested in history and adventure. I also photographed airplanes with my trusty Olympus—but the photos I took were abstracts involving color and reflection, nothing that truly looked like an airplane.

Elsewhere in this book you'll read about my parents building an airplane in our living room. My father will give you the impression that this was a project that the entire family gleefully embraced. Instead of curling up in front of the TV with a bowl of popcorn, he will tell you we gathered around the foam and fiberglass Vari-Eze being constructed in the heart of the house. I have to tell you, I was in my room re-reading *The Hobbit*. The only thing worse than the smell of airplane fuel is the smell of hot foam and epoxy.

My mother was a willing partner in all of this, despite her air-sickness. She'd dealt with car-sickness by taking over the wheel, so she thought learning to fly might help. In case any of you queasy women who love men who love airplanes are thinking this might work for you, let me point out that it didn't work for her, either. She tells the story of the time she took my uncle, an Air Force pilot, up for a flight. On landing, he had to take the stick because she was busy turning green and emptying her guts. She did really enjoy the airplane building project, however. There is nothing she likes better than building something, and the living room has always been an annex to the workshop.

At some point, I was able to escape the world of flight by moving to Colorado to work toward a Master of Fine Arts in poetry. You may begin to notice the theme here: I was the cuckoo in the nest of a family of rabid technologists. I've been dragged, protesting and whining and crying, to every science and technology museum in the U.S. and Canada. Luckily, I learned early on to carry a book—any book—with me.

In my married life, we returned to the familiar world of road trips, the kind where you drive for awhile, then get out and stretch your legs.

And then something dreadful happened. We moved back to Kansas to be near my parents, where my youngest son promptly fell in love with flight. He joined the Civil Air Patrol, took flying lessons, aced ground school while flunking basics like crafts, phys ed, and English, and took up one of my father's more endearing/annoying habits (depending on one's mood at the time).

Yes, when he hears the sound of an airplane overhead, his ears prick up, he squints into the sky, and says things like, "Oh, a Cessna 175. That's got a Lycoming 0-320 B-2A 160-hp, 4-cylinder engine and Cessna's Land-O-Matic tricycle landing gear with a steerable nose-wheel," or some such airplaney-thing.

Meanwhile, I'm saying, "Where?" And "Is it the red one?"

Comic writer Bill Bryson, in his book *Notes from a Small Island,* writes that he believes train spotters have a form of autism called Ausberger's Syndrome. It's that mild autism that allows them to focus on the fact that a certain train car has 68 brass bolts holding the windows in place. Not only that, but the true train spotter can tell you when and where the bolts were manufactured, and why they are different from the other 1,068 bolts manufactured there and used in trains. I submit this is also true of airplane aficionados.

Back in Kansas, I met a childhood friend at a party. His parents are horse people; I've always wanted to be a horse person. My friend commented that he had always been so jealous of my brother and I, because our parents were building an airplane in the living room.

"How cool is that?" he said. "We had to muck out the stables every day, and everything was all about horses. All I wanted to do was fly."

I laughed. "All I wanted to do was ride horses," I said. "If only we'd known, we could have switched parents!"

Now, however, with my father retired from flight, my mother glad to hang up her wings, and my son ready to go out into the world on his own, you'd think I would be free to stick to the ground or the sea.

Instead, I've taken to the air once again, via commercial air travel. At the end of a recent trip to Europe, I spent 36 hours of non-stop travel ... well, that's actually a misnomer. I spent 36 hours between here and there, much of which was spent lounging about in airports with hundreds of other really disgruntled would-be travelers. I tried to convince myself that it wasn't airplane travel that I hated, but airports. Still, I tense up on takeoff, and when the plane touches safely down at the other end of the flight, I find myself thinking, "Right. This is where we burst into flame."

Olive Sullivan

But the real legacy I got from my parents was not fear of flying or distaste for technology. It was the love of travel, the romance of the great beyond. And I've come to accept that, if I want to go to England, Alaska, France, or the Costa del Sol, air travel is the best way to get there. I'd like to spend weeks on a tramp steamer (or on horseback), but realistically speaking, the best, fastest, and cheapest way to get from here to there is still to take to the skies. Darn it all.

A family trip to Oshkosh..

Biographies
The Authors/Editors

Editor and author **Ken Larson** began flying in 1966 and has logged thousands of hours in military and private aircraft. He currently flies business jets. Larson is the author of numerous travel magazine articles, textbook chapters related to international law of the sea, two novels, and one non-fiction book. His work has been featured through various lectures and on radio talk shows.

Editor and author **Tom Holton** served five years in the United States Air Force as a pilot. After the air force, he flew commercial aircraft for the next 30 years at Transamerica Airlines, America West, and Japan Airlines. Tom flew business jets as a Pilot in Command for seven years before retiring in 2009.

Contributors

Paul Derocher spent 27 years in the US Navy, experiencing situations most of us can barely imagine. He flew the P-3 Orion submarine hunter against the Soviet Navy, as well as flying the C-130 Hercules in Antarctica. Paul served several tours of duty at the Pentagon and Defense Intelligence overseas. He was a specialist for the South Asian countries of India, Pakistan, and Afghanistan. After retiring from the navy, Captain Derocher flew the MD-80 and Boeing 737 for American Airlines until forced to retire at age 60. He now flies the twin engine Falcon 2000 business jet.

Marion Hodgson was one of the first women in the United States to train as a military pilot in the Women Airforce Service Pilots (WASP) program. Her book *Winning My Wings: A Woman Airforce Service Pilot in World War II,* tells an exuberant story set in 1943 when she and other WASPs earned their hard-won wings learning to fly everything from open-cockpit primary trainers to P-51 Mustangs, B-26 Marauders, and B-29 Superfortresses. Marion is still active with issues related to women in aviation and works hard to keep the rich memory of the WASPs alive.

Carol Pilon, a professional wingwalker, holds several international records for wingwalking. She is the first and only woman in history to walk on a jet-propelled aircraft, was an integral part of an all-woman team, and executed the world's first and second winter wingwalks. Carol was the first Canadian professional wingwalker, mentored a team abroad, and has trained four different pilots for wingwalk accreditation. She is the first woman and first Canadian to own and operate a wingwalking team in North America—Third Strike Wingwalking. She is also an airframe technician.

Nick Qualantone, a former Marine, became a career Army pilot and flew helicopters in Iraq and fixed wing aircraft in Somalia during the "Blackhawk Down" incident, where Nick flew UN support operations. He piloted Signal Intelligence aircraft on the North Korea border. Nick also flew humanitarian medical relief missions in Central America. Then he was recruited by a federal agency and flew covert operations in South America, including Operation Centre Spike, which took down the notorious drug lord Pablo Escobar in Colombia. After the army, Nick began flying business jets in the USA.

Dr. F. Victor Sullivan has enjoyed a lifelong love affair with flight, dating from his early years in Wichita, Kansas, then the capital of US aviation.

Born in 1931, he suffered a permanent crick in his neck from walking around looking at the sky. As he grew up, the Wichita aviation industry grew also, from small companies like Swallow, Cessna, Beechcraft, and

Stearman, to the huge Boeing complex that included McConnel AFB. Sullivan is a private pilot with a total of over 2,200 hours, an instrument rating, and experience flying SEL aircraft from a J3 Cub to a Mooney Executive Turbo. He served as Education Officer for the Pittsburg Kansas Wing of The Civil Air Patrol in the 1970s, attaining the rank of major. Aircraft ownership included flying club membership in both a Piper Tri-Pacer and a Piper 180 Cherokee. His personally owned BE23 Beechcraft Musketeer 180 carried him over 800 hours east to Washington, DC, north to Ely, Minn., south to Miami, Fla., and west to Tucson, Ariz. Sullivan is dean emeritus of the College of Technology at Pittsburg State University in Pittsburg, Kansas..

Olive L. Sullivan grew up in and around planes and airports because of her father's passion for them. She hated every minute of it (well ... not every minute). Her poetry, short fiction, and creative essays have appeared in magazines and literary journals in the United States, Canada, and Great Britain. A chapbook of poems, *We Start as Water,* was published in 2009. She writes a regular column, articles, and book reviews for regional magazines in the Ozarks. Olive is also an award-winning playwright and screenwriter. Her business, Sullivan Ink, provides editing and coaching services to writers and publishers, and writing services to clients in business and education.

Ken Yamada has over 10 years of experience in the video game industry in roles ranging from business development to original game design. A Georgetown University graduate in 1991, he started as an analyst for Goldman Sachs and in 1996 transitioned into third party accounts at Sony Computer Entertainment (SCE) where he was integral to the successful launch of the Sony PlayStation in North America. Yamada then moved into a senior design role within SCE, which became 989 Studios, a Product Development group within Sony Corporation.

Yamada co-founded BBMF in 2002, a mobile content publisher and distributor based in Japan that bought Atlas Mobile and now produces and distributes hundreds of mobile content worldwide.

In 2004, Yamada co-founded Highway 1 Productions based in San Francisco, CA. He helped design and produce their first hit title, a

$7 million multiplayer online PlayStation 2 and XBOX title called 25-TO-LIFE that launched in February 2006. Yamada also negotiated a multinational $14 million deal for HWY1 with Capcom to manage the development of BIONIC COMMANDO a key next generation title for XBOX 360 and PS3.

Most recently in 2007, Yamada founded 415 Games a casual games publisher also based in San Francisco.

Kevin Kasberg was commissioned in the Navy and began flying after college. He primarily flew H-46 Sea Knight and SH-60B Sea Hawk helicopters—making several deployments on combat support ships, cruisers, frigates, and destroyers during his Navy career. After 22 years of service he transitioned to flying civilian business jets and is presently flying the Gulfstream G-200.

Available from NorlightsPress and fine booksellers everywhere

Toll free: 888-558-4354   **Online:** www.norlightspress.com

Shipping Info: Add $2.95 for first item and $1.00 for each additional item

Name _____

Address _____

Daytime Phone _____

E-mail _____

No. Copies	Title	Price (each)	Total Cost
		Subtotal	
		Shipping	
		Total	

Payment by (circle one):
   Check   Visa   Mastercard   Discover   Am Express

Card number _____ 3 digit code _____

Exp.date _____   Signature _____

Mailing Address:
2721 Tulip Tree Rd.
Nashville, IN 47448

Sign up to receive our catalogue at
www.norlightspress.com

Breinigsville, PA USA
12 May 2010
237859BV00004B/3/P